Cambridge Tracts in Mathematics
and Mathematical Physics

GENERAL EDITORS:
H. BASS, J. F. C. KINGMAN,
F. SMITHIES, J. A. TODD AND C. T. C. WALL

No. 36

RANDOM VARIABLES
AND
PROBABILITY DISTRIBUTIONS

RANDOM VARIABLES
AND
PROBABILITY DISTRIBUTIONS

BY

HARALD CRAMÉR

*Former Chancellor of the Swedish Universities
and Professor in the University of Stockholm*

CAMBRIDGE
AT THE UNIVERSITY PRESS
1970

Published by the Syndics of the Cambridge University Press
Bentley House, 200 Euston Road, London N.W. 1
American Branch: 32 East 57th Street, New York 22, N.Y. 10022

This edition © Cambridge University Press 1970

Library of Congress Catalogue Card Number: 74-92246
Standard Book Number: 521 07685 4

First printed 1937
Second edition 1962
Reprinted 1963
Third edition 1970

First printed at the University Printing House, Cambridge
Reprinted by offset-lithography by John Dickens & Co. Ltd, Northampton

CONTENTS

PREFACE

The Mathematical Theory of Probability has lately become of growing importance owing to the great variety of its applications, and also to its purely mathematical interest. The subject of this tract is the development of the purely mathematical side of the theory, without any reference to the applications. The axiomatic foundations of the theory have been chosen in agreement with the theory given by A. Kolmogoroff in his work *Grundbegriffe der Wahrscheinlichkeitsrechnung*, to which I am greatly indebted. In accordance with this theory, the subject has been treated as a branch of the theory of completely additive set functions. The method principally used has been that of *characteristic functions* (or *Fourier-Stieltjes transforms*).

The limitation of space has made it necessary to restrict the programme somewhat severely. Thus in the first place it has proved necessary to consider exclusively probability distributions in spaces of a finite number of dimensions. With respect to the advanced part of the theory, I have found it convenient to confine myself almost entirely to problems connected with the so-called *Central Limit Theorem* for sums of independent variables, and with some of its generalizations and modifications in various directions. This limitation permits a certain uniformity of method, but obviously a great number of important and interesting problems will remain unmentioned.

My most sincere thanks are due to my friends W. Feller, O. Lundberg and H. Wold for valuable help with the preparation of this work. In particular the constant assistance and criticism of Dr Feller has been very helpful to me.

H. C.

Department of Mathematical Statistics
University of Stockholm
December 1936

PREFACE TO THE SECOND EDITION

This Tract has now been out of print for a number of years. Since there still seems to be some demand for it, the Syndics of the Cambridge University Press have judged it desirable to publish a new edition.

However, owing to the vigorous development of Mathematical Probability Theory since 1937, any attempt to bring the book up to date would have meant rewriting it completely, a task that would have been utterly beyond my possibilities under present conditions. Thus I have had to restrict myself in the main to a number of minor corrections, otherwise leaving the work—including the Bibliography—where it was in 1937.

Besides the minor corrections, most of which are concerned with questions of terminology, there are, in fact, only two major alterations. In the first place, a serious error in the statement and proof of Theorem 11 has been put right. Further, the contents of Chapter IV, § 4, which are fundamental for the theory of asymptotic expansions, etc., developed in Chapter VII, have been revised and simplified. This permits a new formulation of the important Lemma 4, on which the proofs of Theorems 24–26 are based. Finally a brief list of recent works on the subject in the English language has been added.

H.C.

University Chancellor's Office
Stockholm
March 1960

PREFACE TO THE THIRD EDITION

When this Tract was first published in 1937, an important part of it was Chapter VII, containing Liapounoff's classical inequality for the remainder in the Central Limit Theorem, as well as the theory of the related asymptotic expansions. For the Third Edition, this chapter has been partly rewritten, and now brings a proof of the sharper inequality due to Berry and Esseen. Moreover, several minor changes have been made, and the terminology has been somewhat modernized.

H. C.

Djursholm
January 1969

ABBREVIATIONS AND NOTATIONS

Symbol	Signification	Explanation
d.f.	Distribution function	*page* 11
pr.f.	Probability function	11
s.d.	Standard deviation	21
$E(X)$	Mean value (or mathematical expectation) of X	20
$D(X)$	Standard deviation of X	21
c.f.	Characteristic function	24
$F(x) = F_1(x) * F_2(x)$	$F(x) = \int_{-\infty}^{\infty} F_1(x-t)\, dF_2(t)$	37
convergence i.pr.	Convergence in probability	39
$(F(x))^{n*}$	$F(x) * F(x) * \ldots$ (n times)	53

The *union* or *sum* of any finite or enumerable sequence of sets S_1, S_2, ... is denoted by

$$S = S_1 + S_2 + \ldots.$$

The *intersection* or *product* of the sets S_1, S_2, ... is denoted by

$$S = S_1 S_2 \ldots.$$

The inclusion sign $\subset$ is used in relations of the type $S_1 \subset S$ indicating that S_1 is a subset of S, and also in relations of the type $x \subset S$ to express the fact that x is an element of the set S.

FIRST PART

PRINCIPLES

CHAPTER I

INTRODUCTORY REMARKS

1. In the most varied fields of practical and scientific experience, cases occur where certain observations or trials may be repeated a large number of times under similar circumstances. Our attention is then directed to a certain quantity, which may assume different numerical values at successive observations. In many cases each observation yields not only one, but a certain number of quantities, say k, so that generally we may say that the result of each observation is a definite point X in a space of k dimensions ($k \geqslant 1$), while the result of the whole series of observations is a sequence of points: $X_1, X_2, \ldots$.

Thus if we make a series of throws with a given number of dice, we may observe the sum of the points obtained at each throw. We are then concerned with a variable quantity, which may assume every integral value between m and $6m$ (both limits inclusive), where m is the number of dice. On the other hand, in a series of measurements of the state of some physical system, or of the size of certain organs in a number of individuals belonging to the same biological species, each observation furnishes a certain number of numerical values, i.e. a definite point X in a space R of a fixed number of dimensions.

In certain cases, the observed characteristic is only indirectly expressed as a number. Thus if, in a mortality investigation, we observe during one year a large number of persons, we may at each observation (i.e. for each person) note the *number of deaths* which take place during the year, so that in this case the observed quantity assumes the value 0 or 1 according as the corresponding person is alive at the end of the year or not.

In a given class of observations, let R denote the set of points which are *a priori* possible positions of our variable point X, and let S be a sub-set of R. Further, let a series of n observations be made, and count the number ν of those observations, where the following *event* takes place: *the point X determined by the observation belongs to S*. Then the ratio ν/n is called the *frequency* of that event or, as we may shortly put it, *the frequency of the relation (or event)* $X \subset S$. Obviously any such frequency always lies between 0 and 1, both limits inclusive. If $S = S_1 + S_2$, where S_1 and S_2 have no common point, and if ν_1/n and ν_2/n are the frequencies corresponding to S_1 and S_2, we obviously have $\nu = \nu_1 + \nu_2$ and thus

$$(1) \qquad \nu/n = \nu_1/n + \nu_2/n.$$

When we are dealing with such frequencies, a certain peculiar kind of regularity very often presents itself. This regularity may be roughly described by saying that, for any given sub-set S, the frequency of the relation (or event) $X \subset S$ *tends to become more or less constant as n increases*. In certain cases, such as e.g. cases of biological measurements, our observations may be regarded as samples from a very large or even infinite population, so that for indefinitely increasing n the frequency would ultimately reach an ideal value, characteristic of the total population.

It is thus suggested that in cases where the above-mentioned type of regularity appears, we should try to introduce a number $P(S)$ to represent such an ideal value of the frequency ν/n corresponding to the sub-set S. The number $P(S)$ is then called *the probability of the sub-set S, or of the event $X \subset S$*. It follows from (1) that we should obviously choose $P(S)$ such that

$$(2) \qquad P(S_1 + S_2) = P(S_1) + P(S_2)$$

for any two sub-sets S_1 and S_2 of R which have no common point. Further, it is obvious that we should always have $P(S) \geqq 0$ and that for the particular set $S = R$ we should have $P(R) = 1$.

The investigation of *set functions* of the type $P(S)$ and their mutual relations is the object of the Mathematical Theory of Probability. This theory should be considered as a branch of Pure Mathematics, founded on an axiomatic basis, in the same sense as Geometry or Theoretical Mechanics.[1] Once the fundamental conceptions have been introduced and the axioms have been laid down (and in this procedure we are, of course, guided by empirical considerations), the whole body of the theory should be constructed by purely mathematical deductions from the axioms. The practical value of the theory will then have to be tested by experience, just in the same way as a theorem in euclidean geometry, which is intrinsically a purely mathematical proposition, obtains a practical value because experience shows that euclidean geometry really conforms with sufficient accuracy to a large group of empirical facts.

We finally point out that, in order to build a perfectly general mathematical theory of the phenomena encountered in connection with experimental situations of the type considered here, it would be necessary to remove the restriction that R should be a space of a finite number of dimensions. We should then have to regard X as an observed point in some space R of a more general nature. For the purposes of this book we shall, however, restrict ourselves to the case when R has a finite—although in some cases very large—number of dimensions.

2. The axiomatic basis of a theory may, of course, always be constructed in many different ways, and it is well known that, with respect to the foundations of the Theory of Probability, there has been a great diversity of opinions.

The type of statistical regularity indicated above was first observed in connection with ordinary games of. chance with cards, dice, etc., and this gave occasion to the origin and early development of the theory.[2] In every game of this character, all

---

[1] This view seems to have been first explicitly expressed by v. Mises [2].
[2] Cf. Todhunter [1].

the results that are *a priori* possible may be arranged in a finite number of cases which are supposed to be perfectly symmetrical. This led to the famous *principle of equally possible cases* which, after having been more or less tacitly assumed by earlier writers, was explicitly framed by Laplace [1], as the fundamental principle of the whole theory.

During the subsequent discussion of this principle, it has been maintained by various authors that the validity of the principle of equally possible cases is necessarily restricted to the field of games of chance. Attempts have been made[1] to establish the theory on an essentially different basis, the probabilities being directly defined as ideal values of statistical frequencies. The most successful attempt on this line is due to v. Mises [2, 3], who endeavours to reach in this way an axiomatic foundation of the theory in the modern sense.

The fundamental conception of the v. Mises theory is that of a *"Kollektiv"*, by which is meant an unlimited sequence K of similar observations, each furnishing a definite point belonging to an *a priori* given space R of a finite number of dimensions. The first axiom of v. Mises then postulates the existence of the limit

$$(3) \qquad \lim_{n \to \infty} \nu/n = P(S)$$

for every simple sub-set $S \subset R$, while the second axiom requires that the analogous limit should still exist and have the same value $P(S)$ for every sub-sequence K' that can be formed from K according to a rule such that it can always be decided whether the nth observation of K should belong to K' or not, *without knowing the result of this particular observation*. It does, however, seem difficult to give a precise mathematical meaning to the condition printed in italics, and the attempts to express the second axiom in a more rigorous way do not, so far, seem to have reached satisfactory and easily applicable results. Though fully recognizing the value of a system of axioms based on the pro-

---

[1] For the history of these attempts, cf. Keynes [1], chaps. VII–VIII.

perties of statistical frequencies, I think that these difficulties must be considered sufficiently grave to justify, at least for the time being, the choice of a different system.

The underlying idea of the system that will be adopted here may be roughly described in the following simple way: *The probability of an event is a definite number associated with that event; and our axioms have to express the fundamental rules for operations with such numbers.*

Following Kolmogoroff [4], we take as our starting-point the observation made above (cf. (2)) that the probability $P(S)$ may be regarded as an *additive function of the set S*. We shall, in fact, content ourselves by postulating mainly the existence of a function of this type, defined for a certain family of sets S in the k-dimensional space R_k to which our variable point X is restricted. and such that $P(S)$ denotes the probability of the relation $X \subset S$.

Thus the question of the validity of the relation (3) will not at all enter into the mathematical theory. For the *empirical verification* of the theory it will, on the other hand, become a matter of fundamental importance to know if, in a given case, (3) is satisfied with a practically sufficient approximation. Questions of verification and application fall, however, outside the scope of the present work, which will be exclusively concerned with the development of the purely mathematical part of the subject.

3. Before giving the explicit statement of our axioms, it will be convenient to discuss here a few preliminary questions related to the theory of point sets and (generalized) Stieltjes integrals in spaces of a finite number of dimensions.[1]

In the first place, we must define the family F of sets S, for which we shall want our additive set function $P(S)$ to be given. If $X = (\xi_1, ..., \xi_k)$ belongs to the k-dimensional euclidean space

---

[1] Reference may be made to the treatises by Hobson [1], Lebesgue [1] and de la Vallée Poussin [1].

R_k, the family F should obviously contain every k-dimensional interval J defined by inequalities of the form

$$a_i < \xi_i \leqq b_i \quad (i = 1, 2, \ldots, k),$$

as we may always want to know the probability of the relation $X \subset J$. It is also obvious that F should contain every set S constructed by performing on intervals J a finite number of additions, subtractions and multiplications. It is even natural to require that it should be possible to perform these operations an infinite number of times without ever arriving at a set S such that the value of $P(S)$ is not defined. Accordingly, we shall assume that $P(S)$ is defined for all *Borel sets*[1] S of R_k.

Every set which can be constructed from intervals J by applying a finite or infinite number of times the three elementary operations is a Borel set. If S_1, S_2, ... are Borel sets in R_k, this also holds true for the two sets

$$\lim \sup S_n = \lim (S_n + S_{n+1} + \ldots),$$

$$\lim \inf S_n = \lim (S_n S_{n+1} \ldots).$$

If $\lim \sup S_n$ and $\lim \inf S_n$ are identical, we put

$$\lim S_n = \lim \sup S_n = \lim \inf S_n,$$

and thus $\lim S_n$ is also a Borel set. In particular, the sum and product of an infinite sequence of Borel sets are always Borel sets.

If no two of the sets S_i have a common point, it follows from the additive property (2) that

$$P(S_1 + \ldots + S_n) = P(S_1) + \ldots + P(S_n)$$

for every finite n. Since the limit $S_1 + S_2 + \ldots$ always exists and is a Borel set, it is natural to require that this relation should hold even as $n \to \infty$, so that we should have

$$P(S_1 + S_2 + \ldots) = P(S_1) + P(S_2) + \ldots.$$

A set function with this property will be called *completely additive*, and it will be assumed that the function $P(S)$ is of this type.

---

[1] Cf. Hobson [1], I, p. 179; Lebesgue [1], p. 117; de la Vallée Poussin [1], p. 33.

Consider now a real-valued point function $g(X)$, defined for all points $X = (\xi_1, \ldots, \xi_k)$ in R_k. $g(X)$ is said to be *measurable B*[1] if, for all real a and b, the set of points X such that $a < g(X) \leqq b$ is a Borel set. Similarly, a vector function $Y = f(X)$, where $Y = (\eta_1, \ldots, \eta_f)$ belongs to a certain f-dimensional space $\mathfrak{N}_f$, is measurable B if every component η_i, regarded as a function of X, is measurable B. If $\mathfrak{S}$ denotes any Borel set in $\mathfrak{N}_f$, and if S is the set of all points X in R_k such that $f(X) \subset \mathfrak{S}$, then S is also a Borel set. (If $f(X)$ never assumes a value belonging to $\mathfrak{S}$, S is of course the empty set.) If $f_1, f_2, \ldots$ are measurable B, so are $f_1 \pm f_2$, $f_1 f_2$, f_1^{-1}, $\limsup f_n$, $\liminf f_n$ and, in the case of convergence, $\lim f_n$.

All sets of points with which we shall have to deal in the sequel are Borel sets, while all point functions are measurable B. Generally this will not be explicitly mentioned, and should then always be tacitly understood.

A *Lebesgue-Stieltjes integral* with respect to the completely additive set function $P(S)$ is, for every bounded and non-negative $g(X)$ and for every set S, uniquely defined by the postulates

(A) $$\int_{S_1 + S_2} g\, dP = \int_{S_1} g\, dP + \int_{S_2} g\, dP,$$

S_1 and S_2 having no common point, and

(B) $$\int_S (g_1 + g_2)\, dP = \int_S g_1\, dP + \int_S g_2\, dP,$$

(C) $$\int_S g\, dP \geqq 0,$$

(D) $$\int_S 1 . dP = P(S).$$

If g is not bounded, we put $g_M = \min(g, M)$ and define $\int_S g\, dP$ as the limit of $\int_S g_M\, dP$ as $M \to \infty$. If the limit is finite, g is said

---

[1] Cf. Hobson [1], I, p. 563; de la Vallée Poussin [1], p. 34.

to be integrable over S with respect to $P(S)$. The extension to functions g which are not of constant sign is performed by putting

$$2\int_S g\,dP = \int_S (|g|+g)\,dP - \int_S (|g|-g)\,dP.$$

For any g such that $|g| < C$ throughout the set S, we then have the mean value theorem

$$\left|\int_S g\,dP\right| < C\,P(S).$$

Let $g_1, g_2, \ldots$ be a sequence of functions such that for all points of S we have $|g_n| < g$, where g is integrable. Then if $\lim g_n$ exists for every point of S, except possibly for a certain set of points $S_1 \subset S$ such that $P(S_1) = 0$, we have

$$\lim \int_S g_n\,dP = \int_S \lim g_n\,dP.$$

It follows that the theorems on continuity, differentiation and integration with respect to a parameter, etc. which are known from elementary integration theory extend themselves immediately to integrals of the type $\int_S g(X,t)\,dP$, where t is a parameter.

The ordinary theorems on repeated integrals[1] are also easily extended to integrals of the type here considered. In particular we have the following result which will be used in Chapter III. Let $P(S)$ be defined in a two-dimensional space R_2 and such that for every two-dimensional interval J $(a_1 < \xi_1 \leqq b_1, a_2 < \xi_2 \leqq b_2)$ we have $\quad P(J) = P_1(J_1)P_2(J_2),$

where $P_1(S)$ and $P_2(S)$ are completely additive set functions in R_1 while J_i denotes the one-dimensional interval $a_i < \xi_i \leqq b_i$. Then if the function $g_1(\xi_1)g_2(\xi_2)$ is integrable over R_2 with respect to $P(S)$, we have

$$\int_{R_2} g_1(\xi_1)g_2(\xi_2)\,dP = \int_{R_1} g_1(\xi_1)\,dP_1 \int_{R_1} g_2(\xi_2)\,dP_2.$$

---

[1] ·Cf. Hobson [1], I, p. 626; de la Vallée Poussin [1], p. 50.

CHAPTER II

AXIOMS AND PRELIMINARY THEOREMS

1. We now proceed to the explicit statement of our axioms.[1] In accordance with the preceding chapter, we denote by R_k a k-dimensional euclidean space with the variable point $X = (\xi_1, \ldots, \xi_k)$, and we consider the family of all Borel sets S in R_k.

Axiom 1. *To every S corresponds a non-negative number $P(S)$, which is called the probability of the relation (or event) $X \subset S$.*

Axiom 2. *We have $P(R_k) = 1$.*

Axiom 3. *$P(S)$ is a completely additive set function, i.e. we have*

$$P(S_1 + S_2 + \ldots) = P(S_1) + P(S_2) + \ldots,$$

where $S_1, S_2, \ldots$ are Borel sets, no two of which have a common point.

The variable point X is then called a *random variable* (or random point, random vector). The set function $P(S)$ is called the *probability function* of X, and is said to define the *probability distribution* in R_k which is attached to the variable X. It is often convenient to use a concrete interpretation of a probability distribution as a distribution of mass of the total amount 1 over R_k, the quantity of mass allotted to any Borel set S being equal to $P(S)$.

It follows immediately from the axioms that we always have

$$0 \leq P(S) \leq 1,$$

and
$$P(S) + P(S^*) = 1,$$

where S and S^* are complementary sets. Further, if S_1 and S_2 are two sets such that $S_1 \supset S_2$, we have $S_1 = S_2 + (S_1 - S_2)$ and thus

$$(4) \qquad P(S_1) \geq P(S_2).$$

---

[1] The fact that we restrict ourselves here to Borel sets in R_k permits some formal simplification of the system of axioms given by Kolmogoroff [4], and of the immediate conclusions drawn from the axioms.

10   AXIOMS AND PRELIMINARY THEOREMS

Theorem 1. *For any sequence of Borel sets* S_1, S_2, ... *in* R_k, *we have*
$$P(\limsup S_n) \geq \limsup P(S_n),$$
$$P(\liminf S_n) \leq \liminf P(S_n).$$
Hence, if $\lim S_n$ *exists, so does* $\lim P(S_n)$, *and we have*

(5) $$P(\lim S_n) = \lim P(S_n).$$

In order to prove this theorem, we shall first show that (5) holds for any *monotone* sequence $\{S_n\}$. If $\{S_n\}$ is an *increasing* sequence, we may in fact write
$$\lim S_n = S_1 + (S_2 - S_1) + (S_3 - S_2) + \dots,$$
and thus obtain from Axiom 3
$$P(\lim S_n) = P(S_1) + P(S_2 - S_1) + P(S_3 - S_2) + \dots$$
$$= P(S_1) + (P(S_2) - P(S_1)) + (P(S_3) - P(S_2)) + \dots$$
$$= \lim P(S_n).$$

For a *decreasing* sequence $\{S_n\}$, the same thing is shown by considering the increasing sequence formed by the complementary sets S_n^*.

For any sequence $\{S_n\}$, whether monotone or not, we have (cf. I, § 3) $\limsup S_n = \lim (S_n + S_{n+1} + \dots)$. Now, $S_n + S_{n+1} + \dots$ is obviously the general element of a decreasing sequence, so that

(6) $$P(\limsup S_n) = \lim P(S_n + S_{n+1} + \dots).$$

For every $r = 0, 1, \dots$, we have $S_n + S_{n+1} + \dots \supset S_{n+r}$, and thus by (4)
$$P(S_n + S_{n+1} + \dots) \geq P(S_{n+r}),$$
$$P(S_n + S_{n+1} + \dots) \geq \limsup P(S_n).$$

We thus obtain from (6)
$$P(\limsup S_n) \geq \limsup P(S_n).$$

Hence the inequality for $P(\liminf S_n)$ is obtained by considering the sequence $\{S_n^*\}$ of complementary sets and using the identity $\liminf S_n = (\limsup S_n^*)^*$. Thus Theorem 1 is proved.

In the particular case when every point X of R_k belongs at most to a finite number of the sets S_n, $\lim S_n$ is the empty set, and it follows that we have $\lim P(S_n) = 0$.

2. Consider now the particular set $S_{x_1, x_2, \ldots, x_k}$ defined by the inequalities

(7) $$\xi_i \leqslant x_i \quad (i = 1, 2, \ldots, k).$$

For all real values of the x_i we define a *point function* $F(x_1, \ldots, x_k)$ by putting

$$F(x_1, \ldots, x_k) = P(S_{x_1, \ldots, x_k}),$$

so that according to Axiom 1 $F(x_1, \ldots, x_k)$ represents the probability of the joint existence of the relations (7). Then $F(x_1, \ldots, x_k)$ is called the *distribution function*[1] of the probability distribution defined by $P(S)$.

In the sequel, the terms *probability function* and *distribution function* will usually be abbreviated to pr.f. and d.f. respectively.

Let J denote the half-open k-dimensional interval defined by the inequalities $a_i < \xi_i \leqq b_i$ for $i = 1, 2, \ldots, k$. The corresponding probability $P(J)$ is then easily seen to be given by the k-th order difference of the d.f. $F(x_1, \ldots, x_k)$ associated with the interval J. We thus have, writing only the first and last terms of the expression for this difference,

$$P(J) = \Delta_k F(x_1, \ldots, x_k)$$
$$= F(b_1, \ldots, b_k) - \ldots + (-1)^k F(a_1, \ldots, a_k).$$

Theorem 2. *Every d.f.* $F(x_1, \ldots, x_k)$ *possesses the following properties:*

(a) *In each variable* x_i, F *is a never decreasing function, which is everywhere continuous to the right and tends to the limit* 0 *as* $x_i \to -\infty$.

(b) *As all the variables* x_i *tend (independently or not) to* $+\infty$, F *tends to the limit* 1.

(c) *For any half-open* k-*dimensional interval* J, *the associated* k-*th order difference of* F *is non-negative, i.e.* $\Delta_k F \geqq 0$.

Further, every function $F(x_1, \ldots, x_k)$ *which possesses the properties* (a), (b) *and* (c) *determines uniquely a probability distribution in* R_k, *such that* F *represents the probability of the relations* (7).

---

[1] The use here made of the terms *probability function* and *distribution function* corresponds to the terminology of Kolmogoroff [4]. The latter term was used, with the same significance, already by v. Mises [1, 2].

That F is a never decreasing function of each x_i follows immediately from (4), since the set $S_{x_1, ..., x_k}$ increases steadily with each x_i. Further, we have for every $h > 0$

$$F(x_1 + h, x_2, ..., x_k) - F(x_1, x_2, ..., x_k) = P(S_{x_1+h, x_2, ..., x_k} - S_{x_1, x_2, ..., x_k}).$$

If h runs through a sequence of values tending to zero, the sequence of point sets appearing in the second member obviously tends to a definite limit, viz. the empty set. Thus by Theorem 1 the first member tends to zero, and F is continuous to the right in x_1. The same argument evidently applies to every x_i. In the same way it is seen that F tends to zero as any given $x_i \to -\infty$, since the set $S_{x_1, ..., x_k}$ tends then to the empty set.

As, on the other hand, all the variables x_i tend simultaneously to $+\infty$, the set $S_{x_1, ..., x_k}$ tends to the whole space R_k, and consequently F tends to the limit 1.

Further, it is obvious that any d.f. will satisfy the property (c), as we must have $P(S) \geq 0$ for any Borel set S.

The last part of Theorem 2, which asserts that every d.f. uniquely determines a non-negative set function $P(S)$ satisfying our axioms, is equivalent to a well-known proposition in the theory of Lebesgue integration.[1] We have already seen that the d.f. immediately determines the value of $P(S)$ for every half-open k-dimensional interval $a_i < \xi_i \leq b_i$ $(i = 1, 2, ..., k)$. Now every Borel set can be constructed from such intervals by means of repeated passages to the limit, and the corresponding value of the set function has then to be determined according to (5). That this procedure leads to a uniquely determined result for every Borel set is precisely asserted by the proposition referred to above.

According to Theorem 2, we are at liberty to define a probability distribution either by its pr.f. (which is a *set function*) or by its d.f. (which is a *point function*). Though of course the distinction between the two methods is only formal, it will sometimes be found convenient to prefer one of them to the

---

[1] Lebesgue [1], pp. 168–169 (one-dimensional case); de la Vallée Poussin [1], chap. VI.

other. It is particularly in the case of distributions in a one-dimensional space $(k=1)$ that we shall use the d.f., while for general values of k the pr.f. will be used.

In the one-dimensional case $(k=1)$, the property (c) is implied by (a), and thus it follows from Theorem 2 that every non-decreasing function $F(x)$ which is always continuous to the right and is such that $F(x) \to 0$ as $x \to -\infty$, and $F(x) \to 1$ as $x \to +\infty$, defines a probability distribution. As soon as $k>1$, however, (c) is no longer implied by (a), and already for $k=2$ it is in fact easy to construct examples of functions F satisfying (a) and (b), but not (c). Accordingly, these functions are not distribution functions.[2]

3. Let $X=(\xi_1, \ldots, \xi_k)$ be a random variable in R_k with the pr.f. $P(S)$, and $Y=f(X)=(\eta_1, \ldots, \eta_f)$ be a B-measurable function which is finite and uniquely defined for all points X of R_k, and such that its values belong to a certain space $\mathfrak{R}_f$. Then if $\mathfrak{S}$ is a Borel set in $\mathfrak{R}_f$, the set S of all points X in R_k, such that $Y=f(X) \subset \mathfrak{S}$, is (cf. I, §3) also a Borel set. If, now, we define a set function $\mathfrak{P}(\mathfrak{S})$ in $\mathfrak{R}_f$ by the relation

$$\mathfrak{P}(\mathfrak{S}) = P(S),$$

it is readily seen that our Axioms 1–3 are satisfied by $\mathfrak{P}(\mathfrak{S})$, so that $\mathfrak{P}(\mathfrak{S})$ determines a probability distribution in $\mathfrak{R}_f$. *This, by definition, is the probability distribution of the random variable* $Y=f(X)$. The condition that f should be finite and uniquely defined for *all* points of R_k may obviously be replaced by the more general condition that the points X, where f is not finite or not uniquely defined, should form a set Σ such that $P(\Sigma)=0$.

For a set $\mathfrak{S}$ such that the corresponding set S contains no point X we obtain, of course, $\mathfrak{P}(\mathfrak{S})=0$.

Take, e.g., $Y=(\xi_1, \ldots, \xi_f)$, where $f < k$, so that Y is simply the

---

[2] A simple example is the function $F(x, y)$ defined by $F=0$ for $x<1$, $y<1$, for $x \geq 1$, $y<0$, and for $x<0$, $y \geq 1$, and by $F=1$ elsewhere. For this function, the difference $\Delta_2 F$ associated with a sufficiently small interval J containing the point $x=y=1$ in its interior is seen to be negative, so that (c) is not satisfied.

projection of the point X on a certain $\mathfrak{f}$-dimensional sub-space $R_{\mathfrak{f}}$. The pr.f. of Y is then $\mathfrak{P}(\mathfrak{S}) = P(S)$, where S is the *cylinder set* in R_k defined by the relation $(\xi_1, \ldots, \xi_{\mathfrak{f}}) \subset \mathfrak{S}$. This may be concretely interpreted by saying that the distribution of Y is formed by projecting the mass in the original distribution on the sub-space $R_{\mathfrak{f}}$.

In particular $(\mathfrak{f} = 1)$, every component ξ_i of X is itself a random variable, and the corresponding distribution is found by projecting the original distribution on the axis of ξ_i.

4. Two random variables $X_1 = (\xi_1, \ldots, \xi_{k_1})$ in R'_{k_1} and $X_2 = (\eta_1, \ldots, \eta_{k_2})$ in R''_{k_2} being given, it often occurs that we have to consider also the "combined" variable $\mathfrak{X} = (X_1, X_2)$ as a random variable. The "values" of $\mathfrak{X}$ are all pairs of "values" of X_1 and X_2, so that $\mathfrak{X} = (X_1, X_2) = (\xi_1, \ldots, \xi_{k_1}, \eta_1, \ldots, \eta_{k_2})$ is defined in the product space $\mathfrak{R}_{\mathfrak{f}} = R'_{k_1} \cdot R''_{k_2}$, where $\mathfrak{f} = k_1 + k_2$. Obviously the probability distribution of $\mathfrak{X}$ in $\mathfrak{R}$ must be such that its projections on R' and R'' coincide with the distributions of X_1 and X_2 respectively.[1] Similar remarks apply to the "combined" variable formed with any number of random variables.

Let the probability functions of X_1, X_2 and $\mathfrak{X}$ be P_1, P_2 and $\mathfrak{P}$, while the corresponding distribution functions are F_1, F_2 and $\mathfrak{F}$. Then $F_1(x_1, \ldots, x_{k_1})$ and $F_2(y_1, \ldots, y_{k_2})$ denote the probabilities of the relations

$$\xi_i \leqq x_i \quad (i = 1, 2, \ldots, k_1),$$

and

$$\eta_j \leqq y_j \quad (j = 1, 2, \ldots, k_2)$$

respectively, while $\mathfrak{F}(x_1, \ldots, x_{k_1}, y_1, \ldots, y_{k_2})$ denotes the probability of the joint existence of all these $\mathfrak{f} = k_1 + k_2$ relations.

We now introduce the following important definition: The variables X_1 and X_2 are called *mutually independent* if, for all values of the x_i and y_j,

$$(8) \quad \mathfrak{F}(x_1, \ldots, x_{k_1}, y_1, \ldots, y_{k_2}) = F_1(x_1, \ldots, x_{k_1}) F_2(y_1, \ldots, y_{k_2}).$$

If S_1 and S_2 are given sets in R' and R'' respectively, and if we consider the set $\mathfrak{S}$ in $\mathfrak{R}$ formed by all pairs $\mathfrak{X} = (X_1, X_2)$ such that

$$(9) \quad X_1 \subset S_1 \quad \text{and} \quad X_2 \subset S_2,$$

---

[1] Any distribution satisfying this condition is, of course, logically possible.

then it follows from (8) by the basic property of Borel sets that
$$\mathfrak{P}(\mathfrak{S}) = P_1(S_1) P_2(S_2).$$
Thus for two independent variables the probability of the joint existence of the relations (9) is equal to the product of the probabilities for each relation separately. The validity of this multiplicative rule for the particular sets connected with the distribution functions is thus equivalent to the validity of the same rule for all Borel sets.

5. Let X_1, X_2, ..., X_n be random variables with the pr.f.'s $P_1, ..., P_n$ and the d.f.'s $F_1, ..., F_n$ defined in the spaces $R', ..., R^{(n)}$ of any number of dimensions. Consider the combined variable $\mathfrak{X}_n = (X_1, ..., X_n)$ with the pr.f. $\mathfrak{P}_n$ and the d.f. $\mathfrak{F}_n$, defined in the product space $\mathfrak{R}^{(n)}$.

$X_1, ..., X_n$ are then called *mutually independent* if
$$\mathfrak{F}_n = F_1 F_2 ... F_n,$$
which is the straightforward generalization of (8). As in the case of two variables, this is equivalent to the relation
$$(10) \qquad \mathfrak{P}_n(\mathfrak{S}_n) = P_1(S_1) ... P_n(S_n),$$
where $S_1, ..., S_n$ are given sets in $R', ..., R^{(n)}$ respectively, while $\mathfrak{S}_n$ denotes the set in $\mathfrak{R}^{(n)}$ which consists of all points $\mathfrak{X}_n = (X_1, ..., X_n)$ such that $X_i \subset S_i$ for $i = 1, 2, ..., n$.

If, in (10), we put $S_n = R^{(n)}$, we obtain
$$\mathfrak{P}_{n-1}(\mathfrak{S}_{n-1}) = P_1(S_1) ... P_{n-1}(S_{n-1}),$$
where $\mathfrak{P}_{n-1}$ is the pr.f. of $\mathfrak{X}_{n-1} = (X_1, ..., X_{n-1})$ and $\mathfrak{S}_{n-1}$ is the set of all points $\mathfrak{X}_{n-1}$ in $\mathfrak{R}^{(n-1)}$ such that $X_i \subset S_i$ for $i = 1, 2, ..., n-1$. Thus we infer that the variables $X_1, ..., X_{n-1}$ are independent, and in the same way we obviously find that *any group of $m \leqq n$ among the variables X_i are mutually independent.*

Further, it is easily found that, if the variables $X_1, ..., X_m$, $Y_1, ..., Y_n$ are all mutually independent, then the combined variables $\mathfrak{X}_m = (X_1, ..., X_m)$ and $\mathfrak{Y}_n = (Y_1, ..., Y_n)$ are also independent.

6. Any B-measurable vector function $f(X_1, ..., X_m)$ of m random variables may be considered as a B-measurable function of the combined variable $\mathfrak{X}_m$. Thus according to II, § 3, the probability distribution of f is uniquely determined by the distribution of $\mathfrak{X}_m$.

Theorem 3. *Let $X_1, \ldots, X_m, Y_1, \ldots, Y_n$ be independent random variables, and $f(X_1, \ldots, X_m)$ and $g(Y_1, \ldots, Y_n)$ be B-measurable vector functions of the assigned arguments. Then f and g are mutually independent random variables.*

We have $f = \mathfrak{f}(\mathfrak{X}_m)$ and $g = \mathfrak{g}(\mathfrak{Y}_n)$, where, according to the preceding paragraph, $\mathfrak{X}_m$ and $\mathfrak{Y}_n$ are independent. The probability that f belongs to a given set S is then, by definition, equal to the probability that $\mathfrak{X}_m$ belongs to the set of all points which, by the relation $f = \mathfrak{f}(\mathfrak{X}_m)$, correspond to values of f belonging to S. For g, and for the combined variable (f, g), the analogous relations hold. The independence of f and g thus follows from the independence of $\mathfrak{X}_m$ and $\mathfrak{Y}_n$.

7. If X and Y are random variables with the pr.f.'s $P(S)$ and $Q(T)$, where S and T are variable sets in the spaces of X and Y respectively, and if the pr.f. of the combined variable (X, Y) is known, we can form the probability of the joint existence of the relations $X \subset S$, $Y \subset T$. This is a function of two variable sets, say $G(S, T)$. Now let us consider the expression

$$(11) \qquad P_T(S) = \frac{G(S, T)}{Q(T)}$$

for a fixed set T such that $Q(T) > 0$. Then $P_T(S)$ becomes a function of the variable set S, and it is immediately seen that this function satisfies Axioms 1–3. For every fixed T in the space of Y such that $Q(T) > 0$, $P_T(S)$ thus defines a probability distribution in the space of X. This distribution is called *the conditional probability distribution of X relative to the hypothesis $Y \subset T$*, and the quantity $P_T(S)$ is known as *the conditional probability of the relation (event) $X \subset S$ relative to the hypothesis $Y \subset T$*. Similarly, we define a distribution in the space of Y relative to the hypothesis $X \subset S$:

$$(12) \qquad Q_S(T) = \frac{G(S, T)}{P(S)}.$$

for a fixed S such that $P(S) > 0$.

If $S_1, ..., S_n$ are such that S_i and S_k have no common point for $i \neq k$, while $S_1 + ... + S_n$ coincides with the whole space of X, we obviously have by (12)

$$Q(T) = G(S_1, T) + ... + G(S_n, T)$$
$$= P(S_1) Q_{S_1}(T) + ... + P(S_n) Q_{S_n}(T),$$

and so obtain from (11)

$$P_T(S_i) = \frac{P(S_i) Q_{S_i}(T)}{\sum\limits_{i=1}^{n} P(S_i) Q_{S_i}(T)}.$$

This relation is known under the name of *Bayes' theorem*, and is considered as giving the probability *a posteriori*, i.e. calculated after the "event" $Y \subset T$ has been observed, of the particular "hypothesis" $X \subset S_i$, when $P(S_1)$, ..., $P(S_n)$ are the *a priori* probabilities of the various hypotheses $X \subset S_i$.

If X and Y are *independent* variables, we have

$$G(S, T) = P(S) Q(T),$$

and thus by (11) and (12)

$$P_T(S) = P(S), \quad Q_S(T) = Q(T),$$

so that in this case the conditional probabilities coincide with the "total" probabilities $P(S)$ and $Q(T)$.

SECOND PART

DISTRIBUTIONS IN R_1

All random variables and distributions considered in this part are, unless explicitly stated otherwise, defined in a one-dimensional space R_1.

CHAPTER III

GENERAL PROPERTIES. MEAN VALUES

1. According to Theorem 2, the d.f. $F(x)$ of a probability distribution in R_1 is always a non-decreasing function of x, which is everywhere continuous to the right and tends to 0 as $x \to -\infty$, and to 1 as $x \to +\infty$. Conversely, any $F(x)$ with these properties determines a probability distribution.

Any d.f. $F(x)$ being a monotone function, we can at once state a number of general properties of $F(x)$, for the proofs of which we refer to standard treatises on the Theory of Functions of a Real Variable.[1]

Theorem 4. *A d.f. $F(x)$ has at most a finite number of points at which the saltus is $\geq k > 0$, and consequently at most an enumerable set of points of discontinuity. The derivative $F'(x)$ exists for "almost all" values of x (i.e. the points of exception form a set of measure zero).*

$F(x)$ can always be represented as a sum of three components

(13) $$F(x) = a_{\mathrm{I}} F_{\mathrm{I}}(x) + a_{\mathrm{II}} F_{\mathrm{II}}(x) + a_{\mathrm{III}} F_{\mathrm{III}}(x),$$

where a_{I}, a_{II}, a_{III} are non-negative numbers with the sum 1, while F_{I}, F_{II}, F_{III} are distribution functions such that:

$F_{\mathrm{I}}(x)$ is absolutely continuous; $F_{\mathrm{I}}(x) = \displaystyle\int_{-\infty}^{x} F'_{\mathrm{I}}(t)\, dt$ for all values of x.

---

[1] Hobson [1], I, p. 338 and p. 603.

$F_{II}(x)$ *is a "step-function";* $F_{II}(x) = the$ *sum of the saltuses of* $F(x)$ *at all discontinuities* $\leq x$.

$F_{III}(x)$, *the "singular" component, is a continuous function having almost everywhere a derivative* $= 0$.

The three components $a_I F_I$, $a_{II} F_{II}$, $a_{III} F_{III}$ *are uniquely determined by* $F(x)$.

Let us consider in particular the cases when a_I or a_{II} is equal to 1, so that $F(x)$ coincides with $F_I(x)$ or $F_{II}(x)$. We shall say in these cases, which are those usually occurring in the applications, that $F(x)$ is of type I or II respectively.

I. If $F(x) = F_I(x)$, we have for all values of x

$$F(x) = \int_{-\infty}^{x} F'(t) \, dt,$$

and thus the probability that the random variable X with the d.f. $F(x)$ assumes a value belonging to the given set S is

$$\int_S F'(t) \, dt.$$

The derivative $F'(x)$ is then called the *frequency function* or *probability density* of X.

II. If $F(x) = F_{II}(x)$, there is a finite or enumerable set of points x_i such that every x_i is a point of discontinuity of $F(x)$, while $F(x)$ is constant on every closed interval which contains no x_i. If p_i is the saltus of $F(x)$ at the point x_i, we have $\sum_i p_i = 1$.

The probability that X belongs to the given set S is zero, if S does not contain any x_i, and is otherwise equal to the sum of all those p_i which correspond to points x_i belonging to S. Thus in this case the distribution is completely described by saying that we have the probability p_i that X assumes the value x_i $(i = 1, 2, \ldots)$ and the probability 0 that X differs from all the x_i.

2. A Lebesgue-Stieltjes integral $\int g \, dP$ with respect to the pr.f. $P(S)$ has been defined in I, §3. We now define the corresponding integral with respect to the d.f. $F(x)$ simply by putting

$$\int_S g\,dF = \int_S g\,dP.$$

If X is a random variable with the d.f. $F(x)$, the integral $\int_a^b x\,dF(x)$ has a uniquely determined value for all finite a and b. If this integral tends to a finite limit as $a \to -\infty$ and $b \to +\infty$ independently (i.e. if the integral is *absolutely* convergent), we denote this limit by[1]

(14) $$E(X) = \int_{-\infty}^{\infty} x\,dF(x)$$

and call it the *mean value* or *mathematical expectation* of the random variable X.

A B-measurable function $g(X)$ of X may, according to II, § 3, be considered as a random variable. If the d.f. of this variable is denoted by $F^*(x)$, we have by II, § 3,

$$F^*(x) = \int_{S_x} dF(t),$$

where S_x denotes the set of all points t such that $g(t) \leqq x$. Thus we obtain

$$\int_a^b x\,dF^*(x) = \int g(x)\,dF(x),$$

where the integral in the second member is extended to the set $S_b - S_a$. Now if the integral

$$\int_{-\infty}^{\infty} |g(x)|\,dF(x)$$

is convergent, we may allow a and b to tend to $-\infty$ and $+\infty$, and so obtain according to (14) for the *mean value of $g(X)$*

(15) $$E\{g(X)\} = \int_{-\infty}^{\infty} g(x)\,dF(x).$$

In the same way we obtain, if $g(\mathfrak{X})$ is a real-valued function of a random variable $\mathfrak{X}$ which is defined in a space $\mathfrak{R}_t$ of any number

---

[1] In the particular cases when $F(x)$ is of type I or type II, we have

$$E(X) = \int_{-\infty}^{\infty} x F'(x)\,dx \quad \text{and} \quad E(X) = \sum_i p_i x_i$$

respectively.

of dimensions,

(15a) $$E\{g(\mathfrak{X})\}=\int_{\mathfrak{R}_{\mathfrak{f}}}g(\mathfrak{X})\,d\mathfrak{P},$$

where $\mathfrak{P}\,(\mathfrak{S})$ is the pr.f. of $\mathfrak{X}$, and the integral is assumed to be absolutely convergent. If, in particular, $g\,(\mathfrak{X})$ depends only on a certain number $k<\mathfrak{f}$ of the co-ordinates of $\mathfrak{X}$, the integral is, by II, § 3, directly reduced to an integral over the corresponding sub-space $\mathfrak{R}_k$.

The mean value of the particular function $(X-E\,(X))^2$ is called the *variance* of X. The non-negative square root of this mean value is called the *standard deviation* (abbreviated s.d.) of X and is denoted by $D\,(X)$, so that we have, assuming the convergence of the integral.

(16) $$D^2\,(X)=\int_{-\infty}^{\infty}(x-E\,(X))^2\,dF\,(x)$$
$$=E\,(X^2)-E^2\,(X).$$

The square root $D\,(X)$ is always to be given a non-negative value.

We have $D\,(X)=0$ if and only if $F\,(x)$ is constant on every closed interval which does not contain the point $x=E\,(X)$. In this extreme case, we have the probability 1 that the variable X assumes the value $E\,(X)$, and we have $F\,(x)=\epsilon\,(x-E\,(X))$, where $\epsilon\,(x)$ denotes the particular d.f. given by

(17) $$\epsilon\,(x)=\begin{cases}0 \text{ for } x<0,\\1 \ \ ,, \ \ x\geqq 0.\end{cases}$$

In all other cases, the standard deviation $D\,(X)$ is positive.

If X is a random variable with a finite mean value, we obviously have by (15)

(18) $$E\,(aX+b)=aE\,(X)+b$$

for any constant a and b. Further, if the s.d. is also finite, we have

(19) $$D\,(aX+b)=|\,a\,|\,D\,(X).$$

In particular, the *normalized variable* $\dfrac{X-E\,(X)}{D\,(X)}$ has the mean value 0 and the s.d. 1.

The *moments* α_ν and the *absolute moments* β_ν of the variable X

are the mean values of X^ν and $|X|^\nu$ for $\nu = 0, 1, 2, \ldots$:

$$\alpha_\nu = \int_{-\infty}^{\infty} x^\nu dF(x),$$

$$\beta_\nu = \int_{-\infty}^{\infty} |x|^\nu dF(x).$$

(β_ν is, of course, hereby defined also for non-integral $\nu > 0$.) It is immediately seen that, if β_k is finite, both α_ν and β_ν are finite for $\nu \leq k$. Further, we have $\alpha_{2\nu} = \beta_{2\nu}$ and $|\alpha_{2\nu+1}| \leq \beta_{2\nu+1}$. From (14) and (16) we obtain

$$E(X) = \alpha_1, \quad D^2(X) = \alpha_2 - \alpha_1^2.$$

If β_k is finite, it follows from well-known inequalities[1] that we have

(20) $$\beta_1 \leq \beta_2^{\frac{1}{2}} \leq \beta_3^{\frac{1}{3}} \leq \cdots \leq \beta_k^{\frac{1}{k}}.$$

In the sequel it will always be tacitly understood that the mean values occurring in our considerations are assumed to be finite even in the rigorous sense that the corresponding integrals are absolutely convergent.

3. Theorem 5.[2] *Let $\psi(x)$ denote a non-negative function such that $\psi(x) \geq M > 0$ for all x belonging to a certain set S. Then if X is a random variable, the probability that X assumes a value belonging to S is $\leq \dfrac{E\{\psi(X)\}}{M}$.*

This follows directly from the relation

$$E\{\psi(X)\} = \int_{-\infty}^{\infty} \psi(x)\,dF(x) \geq M \int_S dF(x) = MP(S).$$

Taking here in particular $\psi(x) = (x - E(X))^2$, $M = k^2$, we obtain for every $k > 0$ the *Bienaymé-Tchebycheff inequality*: *The probability of the relation $|X - E(X)| \geq k$ is $\leq \dfrac{D^2(X)}{k^2}$.*

Taking further $\psi(x) = |x|^\nu$, $M = k^\nu \beta_\nu$, it follows that the probability of $|X| \geq k \sqrt[\nu]{\beta_\nu}$ is $\leq \dfrac{1}{k^\nu}$.

---

[1] Cf. Hardy-Littlewood-Pólya [1], p. 157.
[2] This is an obvious generalization of theorems due to Tchebycheff and Markoff. Cf. Kolmogoroff [4], p. 37.

Choosing finally $\psi(x) = e^{cx}$, $M = e^{ca}$, where $c > 0$, we conclude that the probability of $X \geqq a$ is $\leqq \dfrac{E(e^{cX})}{e^{ca}}$.

4. Let X and Y be random variables in R_1, such that the combined variable $Z = (X, Y)$ has a certain pr.f. $P(S)$ in R_2. Then $X + Y$ is a one-dimensional vector function of Z, which according to II, §6 has a distribution uniquely determined by $P(S)$. By (15a) we then have $E(X + Y) = \displaystyle\int_{R_2} (X + Y) dP$. The integrals $\displaystyle\int_{R_2} X dP$ and $\displaystyle\int_{R_2} Y dP$ reduce, however, according to the remark made in connection with (15a), to the one-dimensional integrals representing $E(X)$ and $E(Y)$. As soon as these two mean values exist, we thus have the important formula

$$(21) \qquad E(X + Y) = E(X) + E(Y),$$

which is evidently hereby proved without any assumption concerning the nature of the dependence between X and Y. Obviously (21) is immediately generalized to any finite number of terms.

Treating in the same way the product XY, we obtain

$$E(XY) = \int_{R_2} XY \, dP.$$

If, in particular, X and Y are *mutually independent*, we have by II, §4, $P = P_1 P_2$, P_1 and P_2 being the pr.f.'s of X and Y. It then follows from I, §3, that, if X and Y are independent, we have[1]

$$(22) \qquad E(XY) = E(X) E(Y).$$

From (16), (21) and (22) we obtain further, if X and Y are independent,

$$(23) \qquad D^2(X + Y) = D^2(X) + D^2(Y),$$

which is immediately generalized to any finite number of mutually independent terms.

[1] If we restrict ourselves to variables with finite variances, the *necessary and sufficient* condition for the validity of (22) and (23) is that the correlation coefficient of X and Y vanishes.

C

CHAPTER IV

CHARACTERISTIC FUNCTIONS

1. The mean value of a *real* function $g(X)$ of a random variable X has been defined in III, §2. For a *complex* function $g(X) + ih(X)$, we put

$$E(g + ih) = E(g) + iE(h) = \int_{-\infty}^{\infty} (g + ih) \, dF(x).$$

With this definition, the rules for operations with mean values given in the preceding Chapter hold true even for mean values of complex functions. If, in particular, X and Y are complex functions of mutually independent variables, we obtain from (22) and Theorem 3

$$E(XY) = E(X) E(Y).$$

The mean value of the particular function e^{itX}, where t is an auxiliary váriable, will be called the *characteristic function* (abbreviated c.f.) of the corresponding distribution.[1] Denoting this function by $f(t)$, we have

$$(24) \qquad f(t) = E(e^{itX}) = \int_{-\infty}^{\infty} e^{itx} \, dF(x).$$

Unless explicitly stated otherwise, $f(t)$ will be considered for *real values* of t only.

The integral in (24) is absolutely and uniformly convergent for all real t, so that $f(t)$ is uniformly continuous. Obviously $f(0) = 1$, and

$$|f(t)| \leqq \int_{-\infty}^{\infty} dF(x) = 1.$$

The variable $aX + b$ has the c.f. $e^{bit} f(at)$.

---

[1] The first use of an analytical instrument substantially equivalent to the characteristic function seems to be due to Lagrange [1]. (Cf. Todhunter [1], pp. 309–313.) Similar functions were then systematically employed by Laplace in his great work [1].

Thus in particular, putting $E(X) = m$ and $D(X) = \sigma$, the normal-ized variable (cf. III, §2) $\dfrac{X - m}{\sigma}$ has the c.f. $e^{-\frac{mit}{\sigma}} f\left(\dfrac{t}{\sigma}\right)$. Further, the variable $-X$ has the c.f. $f(-t) = \bar{f}(t)$.

Theorem 6.[1] *For every real ξ the limit*

$$\lim_{T \to \infty} \frac{1}{2T} \int_{-T}^{T} f(t) e^{-it\xi} \, dt$$

exists and is equal to the saltus of $F(x)$ at $x = \xi$. Thus if $F(x)$ is continuous at $x = \xi$, the limit is zero.

We have

$$\frac{1}{2T} \int_{-T}^{T} f(t) e^{-it\xi} \, dt = \frac{1}{2T} \int_{-T}^{T} dt \int_{-\infty}^{\infty} e^{it(x-\xi)} \, dF(x)$$

$$= \int_{-\infty}^{\infty} \frac{\sin Tx}{Tx} d_x F(x + \xi).$$

The contribution to the last integral which is due to the domain $|x| \geq h > 0$ tends to zero as $T \to \infty$, whatever the value of h. Let h be so chosen that the variation of $F(x)$ in $(\xi - h, \xi + h)$ exceeds the saltus at $x = \xi$ by less than ϵ. Then, since $\dfrac{\sin Tx}{Tx} = 1$ for $x = 0$, and $\left| \dfrac{\sin Tx}{Tx} \right| \leq 1$ always, it is seen that, for all suffi-ciently large T, the last integral differs from the saltus of F at the point ξ by less than 2ϵ. Thus the theorem is proved.

Representing $F(x)$ as the sum of three components according to (13), we have

$$f(t) = a_\mathrm{I} f_\mathrm{I}(t) + a_\mathrm{II} f_\mathrm{II}(t) + a_\mathrm{III} f_\mathrm{III}(t),$$

each term containing the c.f. of the corresponding component of $F(x)$. We shall consider the behaviour of these three terms separately.

I. Since F_I is absolutely continuous, $f_\mathrm{I}(t) = \displaystyle\int_{-\infty}^{\infty} e^{itx} F_\mathrm{I}'(x) \, dx$,

[1] Bochner [1], p. 79.

and hence $f_{\mathrm{I}}(t) \to 0$ as $|t| \to \infty$, by the Riemann-Lebesgue theorem.[1] It follows that $\dfrac{1}{2T} \displaystyle\int_{-T}^{T} |f_{\mathrm{I}}(t)|^2 dt \to 0$ as $T \to \infty$. If the nth derivative $F_{\mathrm{I}}^{(n)}(x)$ exists for all x and is absolutely integrable, a partial integration shows that $f_{\mathrm{I}}(t) = O\left(\dfrac{1}{|t|^{n-1}}\right)$ as $|t| \to \infty$.

II. $a_{\mathrm{II}} f_{\mathrm{II}}(t) = \sum\limits_{\nu} p_{\nu} e^{itx_{\nu}}$ (using the notations of III, § 1) is the sum of an absolutely convergent trigonometric series, and is thus an almost periodic function,[2] which comes as close to a_{II} as we please for arbitrarily large values of t, so that $\limsup\limits_{|t| \to \infty} |f_{\mathrm{II}}(t)| = 1$. We have[3] $\dfrac{1}{2T} \displaystyle\int_{-T}^{T} |f_{\mathrm{II}}(t)|^2 dt \to \sum\limits_{\nu} p_{\nu}^2/a_{\mathrm{II}}^2$ as $T \to \infty$.

III. $f_{\mathrm{III}}(t)$ is the c.f. of a d.f. $F_{\mathrm{III}}(x)$ which is continuous and has almost everywhere a derivative equal to zero. It is possible to show by examples[4] that $f_{\mathrm{III}}(t)$ does not necessarily tend to 0 as $|t| \to \infty$. We have, however, always $\dfrac{1}{2T} \displaystyle\int_{-T}^{T} |f_{\mathrm{III}}(t)|^2 dt \to 0$ as $T \to \infty$. It will, in fact, be shown in v, § 1, that if $f(t)$ is the c.f. of a *continuous* d.f., the same holds true for $|f(t)|^2$. Thus the desired result follows by applying Theorem 6 to $|f_{\mathrm{III}}(t)|^2$ and putting $\xi = 0$.

We are thus able to state the following theorems.

Theorem 7. *If, in the representation of the d.f. $F(x)$ according to (13), we have $a_{\mathrm{I}} > 0$, then $\limsup\limits_{|t| \to \infty} |f(t)| < 1$.*

If $a_{\mathrm{I}} = 1$, then $\lim\limits_{|t| \to \infty} f(t) = 0$.

If $a_{\mathrm{II}} = 1$, then $\limsup\limits_{|t| \to \infty} |f(t)| = 1$.

Theorem 8.[5] *For every c.f. $f(t)$ we have*

$$\lim_{T \to \infty} \frac{1}{2T} \int_{-T}^{T} |f(t)|^2 dt = \sum_{\nu} p_{\nu}^2,$$

---

[1] Hobson [1], II, p. 514.   [2] Besicovitch [1], p. 6.   [3] Besicovitch [1], p. 19.
[4] Cf. e.g. Jessen-Wintner [1].   [5] Lévy [1], p. 171.

the p_ν being the saltuses of the corresponding d.f. $F(x)$ at all its discontinuities.

Remark. It is easily seen that we cannot have $|f(t_0)| = 1$ for any $t_0 \neq 0$ unless $F(x)$ is of type II, and any two discontinuities x_ν differ by a multiple of $2\pi/t_0$. Hence it follows that, if $\limsup |f(t)| < 1$, then $|f(t)| < k < 1$ for $|t| \geqq \epsilon > 0$, however small ϵ is chosen.

For a later purpose (cf. Theorem 26), we shall in this connection prove the following lemma.

Lemma 1. *If $f(t)$ is a c.f. such that $|f(t)| \leqq k < 1$ as soon as $b \leqq |t| < 2b$, then we have for $|t| < b$*

$$|f(t)| \leqq 1 - (1 - k^2)\frac{t^2}{8b^2}.$$

From the elementary inequality $\cos t \leqq \frac{3}{4} + \frac{1}{4}\cos 2t$ we obtain

$$|f(t)|^2 = \int_{-\infty}^{\infty}\int_{-\infty}^{\infty} e^{it(x-y)}\, dF(x)\, dF(y)$$

$$= \int_{-\infty}^{\infty}\int_{-\infty}^{\infty} \cos t(x-y)\, dF(x)\, dF(y)$$

$$\leqq \tfrac{3}{4} + \tfrac{1}{4}|f(2t)|^2.$$

For $b/2 \leqq |t| < b$ we thus have by hypothesis

$$|f(t)|^2 \leqq 1 - \tfrac{1}{4}(1 - k^2).$$

Repeating the same argument, we conclude that for

$$b/2^n \leqq |t| < b/2^{n-1},$$

where n is an arbitrary integer, we have

$$|f(t)|^2 \leqq 1 - (\tfrac{1}{4})^n(1 - k^2) < 1 - (1 - k^2)t^2/(4b^2),$$

and thus $|f(t)| < 1 - (1 - k^2)t^2/(8b^2).$

As n is arbitrary, this proves our assertion for any t such that $0 < |t| < b$. For $t = 0$, we have $f(0) = 1$, and thus the lemma is proved.

2. If the absolute moment β_k is finite for some positive integer k, (24) may be differentiated k times, and it follows that $f^{(\nu)}(t)$

exists as a bounded and uniformly continuous function for $\nu = 1, 2, \ldots, k$. Obviously $f^{(\nu)}(0) = i^\nu \alpha_\nu$, and so we obtain by MacLaurin's theorem[1] for small values of $|t|$

$$(25) \qquad f(t) = 1 + \sum_1^k \frac{\alpha_\nu}{\nu!} (it)^\nu + o(t^k).$$

For sufficiently small values of $|t|$, the branch of $\log f(t)$ which tends to zero with t may be developed in MacLaurin's series up to the term of order k, and thus we have, introducing a new sequence of parameters,

$$(26) \qquad \log f(t) = \sum_1^k \frac{\gamma_\nu}{\nu!} (it)^\nu + o(t^k).$$

A comparison of (25) and (26) shows that γ_ν is a polynomial in $\alpha_1, \alpha_2, \ldots, \alpha_\nu$ and that $\gamma_1 = \alpha_1$, $\gamma_2 = \alpha_2 - \alpha_1^2$. In the particularly important case $\alpha_1 = 0$, we have

$$\gamma_1 = 0, \ \gamma_2 = \alpha_2, \ \gamma_3 = \alpha_3, \ \gamma_4 = \alpha_4 - 3\alpha_2^2, \ \ldots.$$

The γ_ν are called the *semi-invariants* of the distribution.[2]

For any $n \leq k$, it follows from (25) and (26) that $\gamma_n/n!$ is equal to the coefficient of z^n in the development of $\log\left(1 + \sum_1^n \frac{\alpha_\nu}{\nu!} z^\nu\right)$ as a power series in z. According to (20), this series is majorated by the series

$$-\log\left[1 - \sum_1^\infty \frac{(\beta_n^{\frac{1}{n}} z)^\nu}{\nu!}\right] = \sum_1^\infty \frac{(e^{\beta_n^{\frac{1}{n}} z} - 1)^\nu}{\nu},$$

and so we have

or

$$\frac{|\gamma_n|}{n!} \leq \sum_1^n \left(\text{coeff. of } z^n \text{ in } \frac{1}{\nu} e^{\nu \beta_n^{\frac{1}{n}} z}\right) \leq \frac{n^n \beta_n}{n!}$$

$$(27) \qquad |\gamma_n| \leq n^n \beta_n.$$

3. According to (24), the c.f. $f(t)$ is uniquely determined by the d.f. $F(x)$. We now proceed to prove a group of theorems which show *inter alia* that, conversely, $F(x)$ is uniquely determined by $f(t)$.

---

[1] A form of the remainder in MacLaurin's series which yields (25) does not often occur in text-books. It is, however, easily deduced from the ordinary Lagrange form.
[2] Thiele [1].

Theorem 9.[1] *If $F(x)$ is continuous for $x = \xi$ and for $x = \xi + h$, we have*

$$F(\xi + h) - F(\xi) = \lim_{T \to \infty} \frac{1}{2\pi} \int_{-T}^{T} \frac{1 - e^{-ilh}}{it} e^{-il\xi} f(t)\, dt.$$

Before proving the theorem, we shall use it to prove the identity of any two d.f.'s $F_1(x)$ and $F_2(x)$ having the same c.f. $f(t)$. As a matter of fact, Theorem 9 shows that the differences $F_1(x) - F_1(y)$ and $F_2(x) - F_2(y)$ coincide for almost all values of x and y. If $y \to -\infty$, it follows that $F_1(x) = F_2(x)$ for almost all x. Since every d.f. is continuous to the right, the equality must hold generally.

In order to prove the theorem, we may clearly suppose $h > 0$. Then

$$\frac{1}{2\pi} \int_{-T}^{T} \frac{1 - e^{-ilh}}{it} e^{-il\xi} f(t)\, dt = \int_{-\infty}^{\infty} \psi\, dF(x),$$

where

$$\psi = \psi(x, \xi, h, T) = \frac{1}{\pi} \int_0^T \frac{\sin t\,(x - \xi)}{t}\, dt - \frac{1}{\pi} \int_0^T \frac{\sin t\,(x - \xi - h)}{t}\, dt.$$

Given any $\epsilon > 0$, we now choose δ such that the sum of the variations of $F(x)$ over the intervals $|x - \xi| \leq \delta$ and $|x - \xi - h| \leq \delta$ is less than ϵ. This is possible, since by hypothesis $F(x)$ is continuous for $x = \xi$ and for $x = \xi + h$.

If $T \to \infty$, while ξ and h remain fixed, ψ tends uniformly to 0 in the intervals $x < \xi - \delta$ and $x > \xi + h + \delta$, and to 1 in the interval $\xi + \delta < x < \xi + h - \delta$. In the remaining intervals $|x - \xi| \leq \delta$ and $|x - \xi - h| \leq \delta$, we have $|\psi| < 2$.

It thus follows that, for all sufficiently large values of T, the integral $\int_{-\infty}^{\infty} \psi\, dF(x)$ differs from $\int_{\xi + \delta}^{\xi + h - \delta} dF(x)$ by a quantity of modulus less than 3ϵ. If δ is sufficiently small, the last integral comes, however, as close as we please to $F(\xi + h) - F(\xi)$. Thus the theorem is proved.[2]

[1] Lévy [1], p. 166.

[2] It is easy to show that, if the definition of a d.f. is modified, so that in a point of discontinuity we put $F(x) = \frac{1}{2}[F(x + 0) + F(x - 0)]$, then Theorem 9 holds for *all* values of ξ and h.

The integral appearing in Theorem 9 is, in the general case, only conditionally convergent as $T \to \infty$. We shall now prove a similar theorem which contains an absolutely convergent integral. For any given d.f. $F(x)$ and for any $h > 0$, the function

$$\frac{1}{h}\int_{x}^{x+h} F(u)\,du$$

is obviously a continuous d.f. The corresponding c.f. is found by (24) to be $\dfrac{1-e^{-ith}}{ith} f(t)$. Replacing in Theorem 9 $F(x)$ by this new d.f., we thus obtain for *all* values of ξ and h

$$(28) \quad \frac{1}{h}\int_{\xi+h}^{\xi+2h} F(u)\,du - \frac{1}{h}\int_{\xi}^{\xi+h} F(u)\,du$$
$$= \frac{1}{2\pi h}\int_{-\infty}^{\infty} \left(\frac{1-e^{-ith}}{it}\right)^2 e^{-it\xi} f(t)\,dt.$$

Substituting here ξ for $\xi + h$, we obtain after an easy transformation of the integral on the right-hand side the following theorem.

Theorem 10. *For all real ξ and for all $h > 0$, we have*

$$\frac{1}{h}\int_{\xi}^{\xi+h} F(u)\,du - \frac{1}{h}\int_{\xi-h}^{\xi} F(u)\,du = \frac{1}{\pi}\int_{-\infty}^{\infty} \left(\frac{\sin t}{t}\right)^2 e^{-\frac{2it\xi}{h}} f\left(\frac{2t}{h}\right) dt.$$

This can, of course, also be proved directly, without the use of Theorem 9. We are now in a position to prove the following important theorem.

Theorem 11.[1] *Let $\{F_n(x)\}$ be a sequence of d.f.'s, and $\{f_n(t)\}$ the corresponding sequence of c.f.'s. A necessary and sufficient condition for the convergence of $\{F_n(x)\}$ to a d.f. $F(x)$, in every continuity point of the latter, is that the sequence $\{f_n(t)\}$ of c.f.'s*

---

[1] In a slightly less precise form, this theorem was first proved by Lévy [1], pp. 195–197. Cf. also Bochner [1], p. 72. It should be observed that the theorem becomes false if we omit the assumption that the limit $f(t)$ is continuous at $t = 0$. Choosing, in fact, $f_n(t) = e^{-nt^2}$, we have $f(t) = 1$ for $t = 0$, and $f(t) = 0$ for $t \neq 0$, so that $f(t)$ is discontinuous at $t = 0$. Accordingly, the corresponding sequence of d.f.'s $\{F_n(x)\}$ tends for every x to the limit $F(x) = \frac{1}{2}$, which is not a d.f.

converges for every t to a limit f(t), which is continuous for the special value t = 0.

When this condition is satisfied, the limit f(t) is identical with the c.f. of the limiting d.f. F(x), and f_n(t) converges to f(t) uniformly in every finite t-interval. This implies, in particular, that the limit f(t) is then continuous for all t.

That the condition is *necessary* follows almost immediately from the definition (24) of a c.f. In fact, if $F_n(x) \to F(x)$ in every continuity point of $F(x)$, where $F(x)$ is a d.f., we can choose

$$M = M(\epsilon) \text{ such that } \left| \int_{|x|>M} e^{itx} dF_n(x) \right| < \epsilon \text{ for all } n, \ \epsilon > 0 \text{ being}$$

given. In particular M can be so chosen that $F(x)$ is continuous for $x = \pm M$. According to the theory of Stieltjes integrals, we then have

$$\int_{-M}^{M} e^{itx} dF_n(x) \to \int_{-M}^{M} e^{itx} dF(x),$$

uniformly in every finite t-interval. The last integral differs, however, from the c.f. $f(t)$ of $F(x)$ by a quantity of modulus less than ϵ, if M is sufficiently large. Thus $f_n(t) \to f(t)$ as $n \to \infty$, uniformly in every finite t-interval.

The main difficulty lies in the proof that the condition is *sufficient*. We then assume that $f_n(t)$ tends for every t to a limit $f(t)$ which is continuous for $t = 0$, and we shall prove that under this hypothesis $F_n(x) \to F(x)$ in every continuity point of $F(x)$, where $F(x)$ is a d.f. If this is proved it follows from the first part of the proof that the limit $f(t)$ is identical with the c.f. of $F(x)$, and that $f_n(t)$ converges to $f(t)$ uniformly in every finite t-interval.

In order to prove this, we choose from the sequence $\{F_n(x)\}$ a sub-sequence $F_{n_1}(x), F_{n_2}(x), \ldots$, such that $F_{n_\nu}(x)$ converges to a never decreasing function $F(x)$ in every continuity point of $F(x)$. It is well known that this can always be done, and obviously we may suppose that $F(x)$ is everywhere continuous to the right. We shall now prove that $F(x)$ is a d.f. As we already know that $F(x)$ is never decreasing, and we obviously have

$0 \leq F(x) \leq 1$ for all x, it is sufficient to prove that

$$F(+\infty) - F(-\infty) = 1.$$

From Theorem 10 we obtain, putting $\xi = 0$,

$$(29) \quad \frac{1}{h}\int_0^h F_{n_\nu}(u)\,du - \frac{1}{h}\int_{-h}^0 F_{n_\nu}(u)\,du = \frac{1}{\pi}\int_{-\infty}^\infty \left(\frac{\sin t}{t}\right)^2 f_{n_\nu}\left(\frac{2t}{h}\right)dt.$$

On both sides of this relation, we may allow ν to tend to infinity under the integral signs. In fact, it is easily seen that the convergence conditions for Lebesgue–Stieltjes integrals given in I, §3, are satisfied by the integrals occurring here, and we thus obtain

$$(30) \quad \frac{1}{h}\int_0^h F(u)\,du - \frac{1}{h}\int_{-h}^0 F(u)\,du$$

$$= \frac{1}{\pi}\int_{-\infty}^\infty \left(\frac{\sin t}{t}\right)^2 f\left(\frac{2t}{h}\right)dt.$$

Let now $h \to \infty$. As $F(x)$ is a never decreasing function, the first member of (30) tends to $F(+\infty) - F(-\infty)$. By assumption $f(t)$ is continuous for $t = 0$, so that $f\left(\dfrac{2t}{h}\right)$ tends for every fixed t to the limit $f(0)$. We have, however, $f(0) = \lim_{n \to \infty} f_n(0)$, and $f_n(0) = 1$ for all n, since $f_n(t)$ is a c.f. Hence $f(0) = 1$. Applying once more the convergence properties of integrals (I, §3), we thus obtain

$$F(+\infty) - F(-\infty) = \frac{1}{\pi}\int_{-\infty}^\infty \left(\frac{\sin t}{t}\right)^2 dt = 1.$$

(The value of the last integral may be obtained, e.g. by letting $h \to \infty$ in (29).)

We have thus proved that the sub-sequence $\{F_{n_\nu}(x)\}$ tends to a d.f. $F(x)$, in every continuity point of $F(x)$. By the first part of the proof it then follows that the limit $f(t)$ of the corresponding c.f.'s must be identical with the c.f. of $F(x)$.

Consider now another convergent sub-sequence of $\{F_n(x)\}$, and denote the limit of the new sub-sequence by $F^*(x)$, always assuming this function to be determined so as to be everywhere continuous to the right. In the same way as before, it is then shown that $F^*(x)$ is a d.f. By hypothesis the c.f.'s of the new sub-sequence have, however, for all values of t the same limit $f(t)$ as before, so that $f(t)$ is the c.f. of both $F(x)$ and $F^*(x)$. But then it follows from the remarks made in connection with Theorem 9 that we have $F(x) = F^*(x)$ for all x. Thus every convergent sub-sequence of $\{F_n(x)\}$ has the same limit $F(x)$. This is, however, equivalent to the statement that the sequence $\{F_n(x)\}$ converges to $F(x)$, and since we have shown that $F(x)$ is a d.f., our theorem is proved.

4. Let us now consider a function $R(x)$ which is real and of bounded variation in $(-\infty, +\infty)$, but not necessarily monotone. The integral

$$(31) \qquad r(t) = \int_{-\infty}^{\infty} e^{itx} \, dR(x)$$

is then bounded and uniformly continuous for all real t. In Chapter VII, we shall require the following theorem.

Theorem 12. *Let $R(x)$ be of bounded variation in $(-\infty, +\infty)$, and suppose that $R(x) \to 0$ as $x \to \pm \infty$, so that*

$$(32) \qquad r(0) = \int_{-\infty}^{\infty} dR(x) = 0 \,.$$

Suppose further that we have

$$r(t) = |O(t)$$

as $t \to 0$. For $0 < \omega < 1$, for all real x and all $h > 0$ we then have

$$(33) \qquad \int_{x}^{x+h} (y-x)^{\omega-1} R(y) \, dy$$

$$= -\frac{1}{2\pi i} \int_{-\infty}^{\infty} \frac{r(t)}{t} e^{-itx} \, dt \int_{0}^{h} u^{\omega-1} e^{-itu} \, du \,.$$

If, moreover, the integral

$$(34) \qquad \int_{-\infty}^{\infty} \left| \frac{r(t)}{t} \right| dt$$

is convergent, we have

$$(35) \qquad R(x) = -\frac{1}{2\pi i} \int_{-\infty}^{\infty} \frac{r(t)}{t} e^{-itx} \, dt.$$

We observe that the conditions of the first part of the theorem are satisfied, in particular, whenever $R(x)$ is the difference between two d.f.'s such that the difference between the corresponding c.f.'s is $O(t)$ as $t \to 0$.

In order to prove the theorem, we shall first show that both members of (33) are continuous functions of x and h, when ω is fixed between 0 and 1. In respect of the first member, this is readily seen by writing this member in the form

$$h^{\omega} \int_0^1 y^{\omega-1} R(x+hy) \, dy.$$

In respect of the second member, we have already remarked that $r(t)$ is bounded for all t, and by hypothesis we have $r(t) = O(t)$ as $t \to 0$. Moreover, we have for $t \neq 0$

$$(36) \qquad \left| \int_0^h u^{\omega-1} e^{-itu} \, du \right| = \left| \frac{1}{t^{\omega}} \int_0^{ht} u^{\omega-1} e^{-iu} \, du \right| < \frac{C}{\omega |t|^{\omega}},$$

where C is an absolute constant. It follows that the integral with respect to t in the second member of (33) is absolutely and uniformly convergent for all x and h, and accordingly represents a continuous function.

Without restricting the generality, we may thus assume for the proof of (33) that x and $x+h$ are continuity points of $R(x)$.

The second member of (33) is the limit, as $m \to +0$ and $M \to +\infty$, of the expression

$$-\frac{1}{\pi} \Re \int_m^M \frac{r(t)}{it} e^{-itx} \, dt \int_0^h u^{\omega-1} e^{-itu} \, du$$

$$= -\frac{1}{\pi} \Re \int_0^h u^{\omega-1} \, du \int_m^M \frac{dt}{it} \int_{-\infty}^{\infty} e^{it(y-x-u)} \, dR(y)$$

$$= -\frac{1}{\pi} \int_{-\infty}^{\infty} dR(y) \int_0^h u^{\omega-1} \, du \int_m^M \frac{\sin t\,(y-x-u)}{t} \, dt.$$

According to the convergence properties of integrals (I, § 3), we may here allow m and M to tend to their limits under the integrals. Using the well-known properties of trigonometric integrals, and observing that by assumption we have $R(\pm\infty)=0$, we then find that the second member of (33) is equal to

$$-\frac{1}{\omega}\int_x^{x+h} (y-x)^\omega\, dR\,(y) + \frac{h^\omega}{\omega} R\,(x+h).$$

All integration by parts now yields (33).

On the other hand, replacing the first member of (33) by the expression just obtained, we obtain the relation

$$\int_x^{x+h} (y-x)^\omega\, dR\,(y) - h^\omega R\,(x+h)$$

$$= \frac{\omega}{2\pi i}\int_{-\infty}^{\infty} \frac{r\,(t)}{t}\, e^{-itx}\, dt \int_0^h u^{\omega-1}\, e^{-itu}\, du$$

$$= \frac{1}{2\pi i}\int_{-\infty}^{\infty} \frac{r\,(t)}{t}\, e^{-itx}\, dt \left(h^\omega\, e^{-ith} + it\int_0^h u^\omega\, e^{-itu}\, du \right).$$

If we assume the convergence of (34), this gives (35) as $\omega\to 0$. Thus Theorem 12 is proved.

CHAPTER V

ADDITION OF INDEPENDENT VARIABLES. CONVERGENCE "IN PROBABILITY" SPECIAL DISTRIBUTIONS

1. If X and Y are mutually independent random variables with given d.f.'s $F_1(x)$ and $F_2(y)$, then by II, §4, the d.f. of the combined variable (X, Y) is $F_1(x) F_2(y)$. Thus the pr.f. $\mathfrak{P}(\mathfrak{S})$ of (X, Y) is, according to Theorem 2, uniquely determined by F_1 and F_2 for all two-dimensional Borel sets $\mathfrak{S}$.

The sum $X + Y$ is a one-dimensional vector function of the variable (X, Y), so that according to II, §6, its d.f. $F(z)$ is uniquely determined by $\mathfrak{P}(\mathfrak{S})$, i.e. by F_1 and F_2.

Let $\mathfrak{S}_z$ denote the set of points (X, Y) such that $X + Y \leqq z$. Then by definition $F(z) = \mathfrak{P}(\mathfrak{S}_z)$.

Further, let x_n be a sequence of real numbers steadily increasing with n from $-\infty$ to $+\infty$ and such that $x_{n+1} - x_n < h$ for all n, where h is a given number > 0. Denote by $\mathfrak{R}_n$ the infinite rectangle defined by the inequalities $x_n < X \leqq x_{n+1}$, $Y \leqq z - x_n$, and by $\mathfrak{r}_n$ the rectangle defined by $x_n < X \leqq x_{n+1}$, $Y \leqq z - x_{n+1}$. Obviously $\Sigma \mathfrak{r}_n \subset \mathfrak{S}_z \subset \Sigma \mathfrak{R}_n$, while $\Sigma (\mathfrak{R}_n - \mathfrak{r}_n) \subset \mathfrak{S}_{z+h} - \mathfrak{S}_{z-h}$, the sums being extended from $n = -\infty$ to $n = +\infty$.

Since $\mathfrak{P}(\mathfrak{S})$ is a pr.f., this gives us according to (4)

$$\Sigma \mathfrak{P}(\mathfrak{r}_n) \leqq \mathfrak{P}(\mathfrak{S}_z) \leqq \Sigma \mathfrak{P}(\mathfrak{R}_n)$$

and
$$\Sigma \mathfrak{P}(\mathfrak{R}_n) - \Sigma \mathfrak{P}(\mathfrak{r}_n) \leqq \mathfrak{P}(\mathfrak{S}_{z+h}) - \mathfrak{P}(\mathfrak{S}_{z-h}).$$

The former inequality is equivalent to

$$\Sigma F_2(z - x_{n+1}) (F_1(x_{n+1}) - F_1(x_n)) \leqq F(z)$$
$$\leqq \Sigma F_2(z - x_n) (F_1(x_{n+1}) - F_1(x_n)),$$

while the latter shows that the difference between the limits thus obtained for $F(z)$ does not exceed $F(z+h) - F(z-h)$. In every continuity point of $F(z)$, both sums thus tend to $F(z)$ as $h \to 0$, and by the ordinary definition of a "Riemann-Stieltjes" integral,[1] this limit is equal to $\int_{-\infty}^{\infty} F_2(z-x) dF_1(x)$. According to the definition of a Lebesgue-Stieltjes integral given above (I, §3, and III, §2), the last integral exists, however, for all values of z and is everywhere continuous to the right, so that it always represents $F(z)$. Obviously F_1 and F_2 may be interchanged without altering the value of the integral.

By Theorem 3, any two functions $g_1(X)$ and $g_2(Y)$ are mutually independent, so that we have by (22) $E(g_1 g_2) = E(g_1) E(g_2)$. As pointed out in IV, §1, this holds also if g_1 and g_2 are complex. Thus in particular

$$E(e^{it(X+Y)}) = E(e^{itX}) . E(e^{itY}),$$

so that we have proved the following theorem.

Theorem 13.[2] *If X and Y are mutually independent random variables with the d.f.'s F_1 and F_2, and the c.f.'s f_1 and f_2, then the sum $X + Y$ has the d.f.*

$$(37) \quad F(x) = \int_{-\infty}^{\infty} F_1(x-v) dF_2(v) = \int_{-\infty}^{\infty} F_2(x-v) dF_1(v),$$

and the c.f.

$$(38) \qquad\qquad f(t) = f_1(t) f_2(t).$$

When three d.f.'s satisfy (37), we shall say that F is the *convolution* of the *components* F_1 and F_2, and we shall use the abbreviation

$$(37a) \qquad\qquad F = F_1 * F_2 = F_2 * F_1.$$

According to (38) this symbolical multiplication of the d.f.'s corresponds to a genuine multiplication of the c.f.'s.

If the three variables X_1, X_2 and X_3 are mutually independent, then by Theorem 3 any X_r is independent of the sum of the other

---

[1] Cf. Hobson [1], I, p. 538.

[2] A rigorous proof of this long used theorem, which expresses the fundamental property of the characteristic functions, has not been given until comparatively recently. Cf. Lévy [1], Bochner [1, 2], Wintner [1], Haviland [2].

two variables. A repeated application of Theorem 13 then shows that the sum $X_1 + X_2 + X_3$ has the d.f. $(F_1 * F_2) * F_3 = F_1 * (F_2 * F_3)$, and the c.f. $f_1 f_2 f_3$.

Obviously this may be generalized to any number of components, and it is thus seen that the operation of convolution is commutative and associative. For the sum $X_1 + X_2 + \ldots + X_n$ of n mutually independent variables we have the d.f.

(39) $$F = F_1 * F_2 * \ldots * F_n$$

and the c.f.

(40) $$f = f_1 f_2 \ldots f_n.$$

If at least one of the components F_ν is continuous, it follows from (37) that the convolution F is also continuous.[1] Similarly, if at least one of the F_ν is *absolutely* continuous, this holds also for F. If, on the other hand, all the F_ν have discontinuities, then F has also discontinuities, and the set of discontinuity points of F consists of all points x representable in the form

$$x = x^{(1)} + x^{(2)} + \ldots + x^{(n)},$$

where $x^{(\nu)}$ is a discontinuity point of F_ν.

2. Suppose that the absolute moments of order k are finite for all the mutually independent variables $X_1, X_2, \ldots, X_n$. The inequality $|X_1 + \ldots + X_n|^k \leqq n^{k-1}(|X_1|^k + \ldots + |X_n|^k)$, which holds in every point of the space of the variables $X_1, \ldots, X_n$, then shows that the kth absolute moment of the sum $X_1 + \ldots + X_n$ is also finite.

Further, if $\alpha_1^{(\nu)}, \alpha_2^{(\nu)}, \ldots$ are the moments of F_ν, while $\alpha_1, \alpha_2, \ldots$ are those of the convolution F given by (39), it follows from (25) and (40) that the coefficients of $t, t^2, \ldots, t^k$ in the polynomials

$$1 + \sum_{r=1}^{k} \frac{\alpha_r}{r!} t^r \quad \text{and} \quad \prod_{\nu=1}^{n} \left(1 + \sum_{r=1}^{k} \frac{\alpha_r^{(\nu)}}{r!} t^r \right)$$

are identical. Using a symbolical notation, we may write

(41) $$\alpha_r = (\alpha^{(1)} + \alpha^{(2)} + \ldots + \alpha^{(n)})^r,$$

---

[1] Hence follows the truth of the statement made in IV, § 1: if f is the c.f. of a continuous distribution, then the same holds for $|f|^2 = f \cdot \bar{f}$.

where after the expansion of the rth power, every $(\alpha^{(\nu)})^\mu$ should be replaced by $\alpha_\mu^{(\nu)}$. In particular, we have $\alpha_1 = \Sigma \alpha_1^{(\nu)}$ and $\alpha_2 - \alpha_1^2 = \Sigma \, (\alpha_2^{(\nu)} - (\alpha_1^{(\nu)})^2)$, in accordance with the relations already obtained in III, § 4.

The semi-invariants introduced in IV, § 2, behave in a very simple way when the corresponding probability distributions are convoluted. Let $\gamma_1^{(\nu)}, \dots, \gamma_k^{(\nu)}$ denote the k first semi-invariants of F_ν, which are by hypothesis all finite. Then if $\gamma_1, \dots, \gamma_k$ are the corresponding semi-invariants of F, it follows from (26) and (40) that

$$(42) \qquad\qquad \gamma_r = \sum_{\nu=1}^{n} \gamma_r^{(\nu)}.$$

3. Let $X_1, X_2, \dots$ be a sequence of random variables. We shall say that X_n *converges in probability*[1] (briefly: "converges i.pr.") to a constant A if, for every $\epsilon > 0$, the probability of the relation $|X_n - A| > \epsilon$ tends to zero as $n \to \infty$. We shall also say that X_n converges i.pr. to a random variable X, if the variable $X_n - X$ converges i.pr. to zero.

A proper treatment of the convergence properties of sequences of random variables cannot be given without introducing probability distributions in spaces of an infinite number of dimensions. A few simple theorems will, however, be given here.

A necessary and sufficient condition that X_n converges i.pr. to a constant A is obviously that the d.f. of $X_n - A$ tends, for any $x \neq 0$, to the particular d.f. $\epsilon \, (x)$ defined by (17). By Theorem 11, an equivalent condition is that the corresponding c.f. tends to 1 for all t.

If, for a sequence of variables $Z_1, Z_2, \dots$, the mean value $E \, (Z_n)$ and the s.d. $D \, (Z_n)$ are finite for all n, and if $D \, (Z_n) \to 0$ as $n \to \infty$, it follows immediately from the Bienaymé-Tchebycheff inequality (III, § 3) that $Z_n - E \, (Z_n)$ converges i.pr. to zero. From this

---

[1] Cantelli [1], Slutsky [1], Fréchet [1], Kolmogoroff [4].

D

remark, we deduce at once the following theorem.

Theorem 14. *Let* X_1, X_2, ... *be independent variables such that* $E(X_n) = m_n$ *and* $D(X_n) = \sigma_n$, *and put*

$$Z_n = \frac{1}{n}(X_1 + \ldots + X_n), \quad M_n = \frac{1}{n}(m_1 + \ldots + m_n).$$

If $\sigma_1^2 + \ldots + \sigma_n^2 = o(n^2)$, *then* $Z_n - M_n$ *converges i.pr. to zero.*

We have, in fact, $E(Z_n) = M_n$ and $D^2(Z_n) = \frac{1}{n^2}(\sigma_1^2 + \ldots + \sigma_n^2)$, so that by hypothesis $D(Z_n) \to 0$. In the particular case when all the X_n have the same probability distribution, we have $M_n = m_n = m$, say, and $\sigma_1^2 + \ldots + \sigma_n^2 = n\sigma^2 = o(n^2)$, so that Z_n converges i.pr. to m.

If, for the independent variables X_n considered in Theorem 14, the existence of finite mean values and variances is not assumed, we may still ask if it is possible to find constants M_n such that $\frac{1}{n}(X_1 + \ldots + X_n) - M_n = Z_n - M_n$ converges i.pr. to zero. When all the X_n have the same distribution, the following theorem holds.

Theorem 15.[1] *Let* X_1, X_2, ... *be independent variables all having the same d.f.* $F(x)$, *and put* $Z_n = \frac{1}{n}(X_1 + \ldots + X_n)$. *A necessary and sufficient condition for the existence of a sequence of constants* M_1, M_2, ... *such that* $Z_n - M_n$ *converges i.pr. to zero, is then*

$$\int_{|x|>z} dF(x) = o(1/z)$$

as $z \to \infty$. *This condition being satisfied, we can always take*[2]

$$M_n = \int_{-n}^{n} x \, dF(x).$$

1. *The condition is necessary.* Denoting as usual by $f(t)$ the c.f. corresponding to $F(x)$, the c.f. of $Z_n - M_n$ is

$$e^{-M_n it}\left[f\left(\frac{t}{n}\right)\right]^n = 1 + \lambda_n(t).$$

---

[1] Kolmogoroff [1], and [4], p. 57. Cf. also Khintchine [4].

[2] If, in addition, the generalized mean value $M = \lim\limits_{z \to \infty} \int_{-z}^{z} x \, dF(x)$ exists, it follows that Z_n converges i.pr. to M. If the ordinary mean value as defined in III, §2 exists, it is easily seen that the condition of the theorem is always satisfied.

If $Z_n - M_n$ converges i.pr. to zero, then according to the remark made above the corresponding d.f. tends to $\epsilon(x)$, and thus by Theorem 11 $\lambda_n(t)$ tends to zero, uniformly in every finite t-interval. Taking the n-th root we have therefore as soon as $|\lambda_n(t)| < 1$

$$\left| e^{-\frac{M_n it}{n}} f\left(\frac{t}{n}\right) - 1 \right| = \left| \sqrt[n]{1 + \lambda_n(t)} - 1 \right| \leq \frac{1}{n} \cdot \frac{|\lambda_n(t)|}{1 - |\lambda_n(t)|},$$

while the left side is bounded by 2 for all n and t. Thus

$$f\left(\frac{t}{n}\right) = e^{\frac{M_n it}{n}} + \frac{\theta(n,t)}{n},$$

where $|\theta(n,t)| \leq 2n$ for all n and t, and tends to zero as $n \to \infty$, uniformly in every finite t-interval. Since $f(t/n) \to 1$ as $n \to \infty$, it follows that we have $M_n = o(n)$. From Theorem 10 we then obtain, putting $h = n$,

$$(43) \quad \frac{1}{n} \int_{\xi}^{\xi+n} F(v)\,dv - \frac{1}{n}\int_{\xi-n}^{\xi} F(v)\,dv$$
$$= \frac{1}{\pi}\int_{-\infty}^{\infty} \left(\frac{\sin t}{t}\right)^2 e^{-\frac{2it\xi}{n}}\left(e^{\frac{2M_n it}{n}} + \frac{\theta(n, 2t)}{n}\right) dt.$$

Now, $e^{M_n it}$ is the c.f. of the d.f. $\epsilon(x - M_n)$, where $\epsilon(x)$ is defined by (17). The contribution to the second member of (43) arising from the term $e^{\frac{2M_n it}{n}}$ is thus by Theorem 10 equal to the value assumed by the first member if $F(v)$ is replaced by $\epsilon(v - M_n)$. This value is, however, equal to zero if $|\xi - M_n| > n$. For all ξ satisfying this condition, we thus obtain, since $\theta(n,t)$ tends uniformly to zero,

$$\frac{1}{n}\int_{\xi}^{\xi+n} F(v)\,dv - \frac{1}{n}\int_{\xi-n}^{\xi} F(v)\,dv < \frac{\eta(n)}{n},$$

where $\eta(n) \to 0$ as $n \to \infty$. On the other hand we have

$$\frac{1}{n}\int_{\xi}^{\xi+n} F(v)\,dv - \frac{1}{n}\int_{\xi-n}^{\xi} F(v)\,dv = \int_{\xi-n}^{\xi+n}\left(1 - \frac{|v-\xi|}{n}\right)dF(v)$$
$$\geq \tfrac{1}{2}\left(F(\xi + \tfrac{1}{2}n) - F(\xi - \tfrac{1}{2}n)\right),$$

and thus $$F\left(\xi+\tfrac{1}{2}n\right)-F\left(\xi-\tfrac{1}{2}n\right)<\frac{2\eta\left(n\right)}{n}$$

for all ξ such that $|\xi-M_n|>n$. Since $M_n=o\left(n\right)$, we may for all sufficiently large n put $\xi=\pm\tfrac{3}{2}n$, so that we obtain

$$F\left(2n\right)-F\left(n\right)<\frac{2\eta\left(n\right)}{n}\quad\text{and}\quad F\left(-n\right)-F\left(-2n\right)<\frac{2\eta\left(n\right)}{n}.$$

Replacing here successively n by $2n$, 2^2n, ... and adding, we obtain the desired result, as the restriction of n to integral values is obviously not essential.

2. *The condition is sufficient.* Taking $M_n=\displaystyle\int_{-n}^{n}x\,dF\left(x\right)$, we have by a partial integration

$$|M_n|\leqq\int_{-n}^{n}|x|\,dF\left(x\right)=-n\int_{|v|>n}dF\left(v\right)+\int_{0}^{n}dx\int_{|v|>x}dF\left(v\right)$$

$$=o\left(1\right)+o\left(\int_{0}^{n}\frac{dx}{x+1}\right)=o\left(\log n\right),$$

and in the same way

$$\int_{-n}^{n}x^2dF\left(x\right)=-n^2\int_{|v|>n}dF\left(v\right)+2\int_{0}^{n}x\,dx\int_{|v|>x}dF\left(v\right)=o\left(n\right).$$

The c.f. of Z_n-M_n may be written

$$(44)\quad e^{-M_nit}\left[f\left(\frac{t}{n}\right)\right]^n=\left[1+\int_{-\infty}^{\infty}(e^{\frac{it(x-M_n)}{n}}-1)\,dF\left(x\right)\right]^n.$$

Now we have by hypothesis

$$\int_{-\infty}^{\infty}(e^{\frac{it(x-M_n)}{n}}-1)\,dF\left(x\right)=\int_{-n}^{n}(e^{\frac{it(x-M_n)}{n}}-1)\,dF\left(x\right)+\frac{\theta\left(n,t\right)}{n},$$

where $\theta\left(n,t\right)$ tends to zero as $n\to\infty$, uniformly in every finite t-interval. According to the definition of M_n we may thus write

$$(45)\quad n\int_{-\infty}^{\infty}(e^{\frac{it(x-M_n)}{n}}-1)\,dF\left(x\right)$$

$$=n\int_{-n}^{n}\left(e^{\frac{it(x-M_n)}{n}}-1-\frac{it\left(x-M_n\right)}{n}\right)dF\left(x\right)+\theta\left(n,t\right).$$

The first term in the second member of (45) is, however, of modulus less than

$$\frac{t^2}{2n} \int_{-n}^{n} (x - M_n)^2 \, dF(x),$$

and according to the above inequalities this tends to zero as $n \to \infty$, uniformly in every finite t-interval. Since $(1 + \alpha_n)^n \to 1$ if $n\alpha_n \to 0$, it thus follows from (44) and (45) that the c.f. of $Z_n - M_n$ tends to 1 as $n \to \infty$, uniformly in every finite t-interval. Then by Theorem 11 $Z_n - M_n$ converges i.pr. to zero.

4. Let $X_1, X_2, \ldots$ be independent variables, and put

$$Y_n = X_1 + X_2 + \ldots + X_n..$$

If $F_\nu(x)$ is the d.f. and $f_\nu(t)$ the c.f. of the variable X_ν, the d.f. of Y_n is $F_1 * F_2 * \ldots * F_n$, and the corresponding c.f. is $f_1 f_2 \ldots f_n$. By Theorem 11, a necessary and sufficient condition for the convergence of $F_1 * F_2 * \ldots * F_n$ to a d.f. $F(x)$ is the convergence of the infinite product

$$f(t) = \prod_{\nu=1}^{\infty} f_\nu(t),$$

for all t, where $f(t)$ is continuous for $t = 0$.[1] If this condition is satisfied, it follows from Theorem 11 that the infinite product converges even uniformly in every finite t-interval, and that $f(t)$ is the c.f. of a d.f. $F(x)$ such that

$$F(x) = \lim_{n \to \infty} F_1 * F_2 * \ldots * F_n$$

in every continuity point of F. For any n', the product $\prod_{n+1}^{n+n'} f_\nu(t)$ tends uniformly to 1 as $n \to \infty$, and consequently the difference $Y_{n+n'} - Y_n$ converges i.pr. to zero. It would be natural to conclude that there is a variable Y with the d.f. $F(x)$, such that Y_n converges i.pr. to Y. Then Y would be the sum of an infinite series of random variables: $Y = X_1 + X_2 + \ldots$. In order to give a precise meaning to a statement of this character it is necessary to consider

---

[1] A *sufficient* condition for this convergence is the convergence of the two series $\Sigma E(X_\nu)$ and $\Sigma D^2(X_\nu)$.

a probability distribution in the space of the combined variable $(X_1, X_2, \ldots)$, which has an infinite number of dimensions. This falls, however, outside the scope of the present work.

5. We shall now consider some particular examples of probability distributions. In the first place, we consider a variable X which can assume only the values 1 and 0, the corresponding probabilities being p and $q = 1 - p$. The d.f. of this variable is a "step-function" with steps in the points 1 and 0, of the height p and q respectively, while the corresponding c.f. is equal to $pe^{it} + q$. We have further $E(X) = p$ and $D(X) = \sqrt{pq}$. If X_1, $X_2, \ldots, X_n$ are independent variables all having the same distribution as X, the sum $\nu = X_1 + X_2 + \ldots + X_n$ is equal to the number of those X_r which assume the value 1. The c.f. of the variable ν is $(pe^{it} + q)^n$, and ν may assume the values 0, 1, $\ldots$, n, the probability of a given value ν being $\binom{n}{\nu} p^\nu q^{n-\nu}$. This distribution is usually called a *binomial* or *Bernoulli distribution*, and ν may be concretely interpreted as the number of white balls obtained in a set of n drawings from an urn, the probability of drawing a white ball being each time equal to p. By III, § 4, we have $E(\nu) = np$ and $D(\nu) = \sqrt{npq}$. According to Theorem 14, the "frequency" ν/n converges i.pr. to p. This result coincides with the classical *Bernoulli's theorem* as originally proved by Bernoulli.

If we allow the quantity p to vary from one X_r to another, the c.f. of the sum ν becomes

$$(46) \qquad \prod_1^n (p_r e^{it} + q_r) = \prod_1^n (1 + p_r (e^{it} - 1)).$$

In this case, we have $E(\nu) = \sum_1^n p_r$ and $D(\nu) = \sqrt{\sum_1^n p_r q_r}$, and by Theorem 14 the variable $(\nu - \sum_1^n p_r)/n$ converges i.pr. to zero. If the series $\sum_1^\infty p_r$ is convergent, it is seen that the c.f. (46) tends to a limit as $n \to \infty$, uniformly for all real t, so that the case considered at the end of the preceding paragraph presents itself.

Another case of convergence is obtained if, in (46), we allow the p_r to depend on n in such a way that, when $n \to \infty$, each p_r tends to zero, while $\sum\limits_{1}^{n} p_r$ tends to a constant $\lambda > 0$. (We may e.g. take $p_r = \lambda/n$ for $r = 1, 2, \ldots, n$.) Then the c.f. (46) tends to the limit

$$(47) \qquad e^{\lambda(e^{it}-1)} = \sum_{\nu=0}^{\infty} \frac{\lambda^{\nu}}{\nu!} e^{-\lambda} \cdot e^{\nu it}.$$

This is the c.f. of a variable which may assume the values $0, 1, 2, \ldots$, the probability of any given ν being $\dfrac{\lambda^{\nu}}{\nu!} e^{-\lambda}$. The mean value of this variable is λ, and the s.d. is $\sqrt{\lambda}$. The semi-invariants γ_{μ} defined by (26) are all equal to λ. This distribution is usually called a *Poisson distribution*. If X_1 and X_2 are independent variables both having Poisson distributions with the parameter values λ_1 and λ_2, the expression (47) of the c.f. shows that the sum $X_1 + X_2$ has a distribution of the same kind with the parameter $\lambda_1 + \lambda_2$. If we denote by $F(x, \lambda)$ the d.f. corresponding to the Poisson distribution, we thus have the relation

$$(48) \qquad F(x, \lambda_1) * F(x, \lambda_2) = F(x, \lambda_1 + \lambda_2).$$

6. The probability distribution defined by the d.f. $F(x/a)$, where $a > 0$ and

$$F(x) = \tfrac{1}{2} + \frac{1}{\pi} \arctan x, \quad F'(x) = \frac{1}{\pi} \cdot \frac{1}{1 + x^2},$$

is sometimes called *Cauchy's distribution*.[1] This distribution has not a finite mean value, since the integral $\displaystyle\int_{-\infty}^{\infty} \frac{|x|\, dx}{1 + x^2}$ does not converge, although the "generalized mean value" $\displaystyle\lim_{z \to \infty} \int_{-z}^{z} x\, dF(x)$ does exist and is equal to zero. By an easy application of Cauchy's

---

[1] Cf. Lévy [1], p. 179.

theorem we find the c.f.

$$f(t) = \frac{1}{\pi} \int_{-\infty}^{\infty} \frac{e^{itx}}{1+x^2} dx = e^{-|t|}.$$

The c.f. corresponding to the d.f. $F(x/a)$ is then obviously $f(at) = e^{-a|t|}$. We thus have

$$f(a_1 t) f(a_2 t) = f((a_1 + a_2) t),$$

or
$$F(x/a_1) * F(x/a_2) = F(x/(a_1 + a_2)),$$

so that the Cauchy distribution reproduces itself at the addition of independent variables. If $X_1, X_2, \ldots, X_n$ are independent variables all having the Cauchy d.f. $F(x/a)$, the arithmetic mean $Z_n = (X_1 + \ldots + X_n)/n$ thus has the same d.f. $F(x/a)$. Hence we cannot in this case find constants M_n such that $Z_n - M_n$ converges i.pr. to zero. It is easily seen that, accordingly, the condition of Theorem 15 is not satisfied.

As our next example we take a d.f. $F(x; \alpha, \lambda)$ which is equal to zero for $x \leq 0$, and for $x > 0$ is defined by

$$F(x; \alpha, \lambda) = \frac{\alpha^\lambda}{\Gamma(\lambda)} \int_0^x v^{\lambda-1} e^{-\alpha v} dv, \quad (\alpha > 0, \lambda > 0).$$

with the frequency function

(49)
$$F'(x; \alpha, \lambda) = \frac{\alpha^\lambda}{\Gamma(\lambda)} x^{\lambda-1} e^{-\alpha x}.$$

This is a distribution of "type III" according to the classification introduced by K. Pearson.[1] All moments of the distribution are finite; the mean value is λ/α, and the s.d. is $\sqrt{\lambda}/\alpha$. The c.f. is

$$f(t; \alpha, \lambda) = \frac{\alpha^\lambda}{\Gamma(\lambda)} \int_0^\infty x^{\lambda-1} e^{-(\alpha-it)x} dx = \frac{1}{\left(1 - \dfrac{it}{\alpha}\right)^\lambda}.$$

This shows that we have the expression $\gamma_\mu = (\mu - 1)! \, \alpha^{-\mu} \lambda$ for the semi-invariant γ_μ of the distribution, and that, for a fixed value of α, the d.f. satisfies with respect to the parameter λ the same relation (48) as the Poisson distribution.

---

[1] Cf. e.g. Elderton [1].

In the particular case when $\alpha = \frac{1}{2}$ and $\lambda = n/2$, where n is an integer, the expression (49) of the frequency function becomes

$$(49a) \qquad F'(x; \tfrac{1}{2}, n/2) = \frac{1}{2^{n/2}\,\Gamma(n/2)}\, x^{n/2-1}\, e^{-x/2}.$$

The corresponding distribution has important applications in mathematical statistics, and is known as the χ^2 distribution. Anticipating the discussion of the normal distribution that will be given in the following Chapter VI, we mention that $(49a)$ is the frequency function of the sum of the squares of n independent random variables, each of which has the normal d.f. $\Phi(x)$.

7. In many applications, it is required to find the distribution of the *quotient* of two random variables. In certain cases, the following theorem enables us to express this distribution in terms of the c.f.'s of the two variables.

Theorem 16. *Let X_1 and X_2 be independent variables with finite mean values, the corresponding d.f.'s being $F_1(x)$ and $F_2(x)$, with the c.f.'s $f_1(t)$ and $f_2(t)$. If $F_2(0) = 0$, and if the integral*

$$\int_1^\infty \left| \frac{f_2(t)}{t} \right| dt$$

converges, then the d.f. $G(x)$ of the quotient X_1/X_2 is given by the relation

$$G(x) = \frac{1}{2\pi i} \int_{-\infty}^{\infty} \frac{f_2(t) - f_1(t) f_2(-tx)}{t}\, dt.$$

If the integral obtained by formal differentiation of this relation with respect to x is uniformly convergent in a certain interval, we thus have in this interval for the frequency function $G'(x)$

$$G'(x) = \frac{1}{2\pi i} \int_{-\infty}^{\infty} f_1(t) f_2'(-tx)\, dt.$$

By definition $G(x)$ is equal to the probability of the relation $X_1/X_2 \le x$. Since $F_2(0) = 0$, we need only consider positive values of X_2, so that the last inequality is equivalent to $X_1 - xX_2 \le 0$, and if $H(\xi)$ denotes the d.f. of the variable $X_1 - xX_2$, we thus obtain $G(x) = H(0) = H(0) - F_2(0)$. By hypothesis the difference $H(\xi) - F_2(\xi)$ satisfies the conditions of the last part of Theorem 12, and so we obtain, since the c.f. of $H(\xi)$ is $f_1(t)f_2(-tx)$,

$$H(\xi) - F_2(\xi) = \frac{1}{2\pi i}\int_{-\infty}^{\infty} \frac{e^{-it\xi}}{t} (f_2(t) - f_1(t)f_2(-tx))\,dt.$$

Putting here $\xi = 0$, we obtain the theorem.

We shall give two examples for the application of this theorem. In the first place, we consider two variables X_1 and X_2, both distributed according to (49) with the parameters α_1, λ_1 and α_2, λ_2 respectively. In this case the theorem gives

$$G(x) = \frac{1}{2\pi i}\int_{-\infty}^{\infty} \left(\frac{1}{\left(1 - \dfrac{it}{\alpha_2}\right)^{\lambda_2}} - \frac{1}{\left(1 - \dfrac{it}{\alpha_1}\right)^{\lambda_1}\left(1 + \dfrac{itx}{\alpha_2}\right)^{\lambda_2}} \right) \frac{dt}{t}.$$

If λ_2 is an integer, the integral may be calculated, and we find by an easy application of Cauchy's theorem $G(x) = 0$ for $x \leqq 0$ and

$$G(x) = \left(\frac{\alpha_1 x}{\alpha_1 x + \alpha_2}\right)^{\lambda_1} \sum_{\nu=0}^{\lambda_2-1} \binom{\lambda_1 + \nu - 1}{\nu} \left(\frac{\alpha_2}{\alpha_1 x + \alpha_2}\right)^{\nu}$$

for $x > 0$. In the particular case $\lambda_2 = 1$, the last expression reduces to $G(x) = \left(\dfrac{\alpha_1 x}{\alpha_1 x + \alpha_2}\right)^{\lambda_1}$.

For our second example, we shall again anticipate the discussion of the normal distribution. We shall consider a quotient of the form $X_1/\sqrt{X_2}$, where X_1 is normally distributed with the mean value 0 and the s.d. σ, while X_2 is distributed according to (49). We then have (cf. (51))

$$f_1(t) = e^{-\frac{1}{2}\sigma^2 t^2}$$

and

$$f_2(t) = \frac{\alpha^\lambda}{\Gamma(\lambda)}\int_0^\infty x^{\lambda-1} e^{-\alpha x + it\sqrt{x}}\,dx = \frac{2\alpha^\lambda}{\Gamma(\lambda)}\int_0^\infty v^{2\lambda-1} e^{-\alpha v^2 + itv}\,dv,$$

$$f_2'(t) = \frac{2i\alpha^\lambda}{\Gamma(\lambda)}\int_0^\infty v^{2\lambda} e^{-\alpha v^2 + itv}\,dv.$$

In this case we may apply the last formula of Theorem 16, and so obtain for the frequency function $G'(x)$ of the variable $X_1/\sqrt{X_2}$

$$G'(x) = \frac{\alpha^\lambda}{\pi\Gamma(\lambda)} \int_{-\infty}^{\infty} e^{-\frac{1}{2}\sigma^2 t^2}\, dt \int_0^{\infty} v^{2\lambda}\, e^{-\alpha v^2 - itvx}\, dv$$

$$= \frac{\alpha^\lambda}{\pi\Gamma(\lambda)} \int_0^{\infty} v^{2\lambda}\, e^{-\alpha v^2}\, dv \int_{-\infty}^{\infty} e^{-\frac{1}{2}\sigma^2 t^2 - itvx}\, dt$$

$$= \frac{2\alpha^\lambda}{\sigma\sqrt{2\pi}\,\Gamma(\lambda)} \int_0^{\infty} v^{2\lambda}\, e^{-\left(\alpha + \frac{x^2}{2\sigma^2}\right) v^2}\, dv$$

$$= \frac{1}{\sqrt{2\pi\alpha\sigma^2}} \frac{\Gamma(\lambda+\frac{1}{2})}{\Gamma(\lambda)} \left(1 + \frac{x^2}{2\alpha\sigma^2}\right)^{-\lambda-\frac{1}{2}}.$$

This is a distribution of type VII according to the classification of K. Pearson. In the particular case when $\sigma^2 = n$, while X_2 has the χ^2 distribution given by (49a), the expression of $G'(x)$ becomes

$$G'(x) = \frac{1}{\sqrt{n\pi}} \frac{\Gamma\left(\dfrac{n+1}{2}\right)}{\Gamma\left(\dfrac{n}{2}\right)} \left(1 + \frac{x^2}{n}\right)^{-\frac{n+1}{2}},$$

which is known under the name of "*Student's*" distribution.[1]

---

[1] "Student" [1].

CHAPTER VI

THE NORMAL DISTRIBUTION AND THE CENTRAL LIMIT THEOREM

1. The *normal distribution function*[1] $\Phi(x)$ is defined by the relation

$$\Phi(x) = \frac{1}{\sqrt{2\pi}} \int_{-\infty}^{x} e^{-\frac{t^2}{2}} dt.$$

The corresponding *normal frequency function* is

$$\Phi'(x) = \frac{1}{\sqrt{2\pi}} e^{-\frac{x^2}{2}}.$$

The mean value of this distribution is 0, and the s.d. is 1, as shown by the relations

$$(50) \qquad \int_{-\infty}^{\infty} x\, d\Phi(x) = 0, \qquad \int_{-\infty}^{\infty} x^2\, d\Phi(x) = 1.$$

The moments of odd order $\alpha_{2\nu+1}$ all vanish, while

$$\alpha_{2\nu} = \int_{-\infty}^{\infty} x^{2\nu}\, d\Phi(x) = 1.3.\dots.(2\nu-1).$$

The c.f. is, by a well-known integral formula,

$$(51) \qquad \int_{-\infty}^{\infty} e^{itx}\, d\Phi(x) = \frac{1}{\sqrt{2\pi}} \int_{-\infty}^{\infty} e^{itx-\frac{x^2}{2}} dx = e^{-\frac{t^2}{2}}.$$

Hence we obtain, for $\nu = 1, 2, \dots$, by partial integration

$$(52) \qquad \int_{-\infty}^{\infty} e^{itx}\, d\Phi^{(\nu)}(x) = (-it)^{\nu} e^{-\frac{t^2}{2}},$$

and by differentiation

$$\Phi^{(\nu)}(x) = \frac{1}{2\pi} \int_{-\infty}^{\infty} (it)^{\nu-1} e^{itx-\frac{t^2}{2}} dt.$$

A random variable X is said to be *normally distributed*, if its

---

[1] The normal distribution was discussed already in 1733 by De Moivre in the second supplement to his *Miscellanea Analytica*. Cf. K. Pearson [**1**]. It was afterwards treated by Gauss and Laplace, and is often referred to as the Gauss or Gauss-Laplace distribution.

d.f. is $\Phi\left(\dfrac{x-m}{\sigma}\right)$, where $\sigma \geqq 0$ and m are constants. (The case $\sigma = 0$ is, of course, a degenerated limiting case which might be called an *improper* normal distribution. $\Phi\left(\dfrac{x-m}{0}\right)$ should always be interpreted as $\epsilon\,(x-m)$, where $\epsilon\,(x)$ is defined by (17).) The *normalized* variable $\dfrac{X-m}{\sigma}$ has then the d.f. $\Phi\,(x)$, and we obtain from (50)

$$E\,(X) = m, \quad D\,(X) = \sigma,$$

while (51) shows (cf. also IV, § 1) that the c.f. of the variable X is

$$E\,(e^{itX}) = e^{mit - \frac{1}{2}\sigma^2 t^2}.$$

The semi-invariants of X, as defined in IV, § 2, are

$$\gamma_1 = m, \quad \gamma_2 = \sigma^2, \quad \gamma_3 = \gamma_4 = \ldots = 0.$$

2. We now proceed to prove a number of theorems which show that the normal distribution plays a fundamental part in a great number of questions connected with the addition of mutually independent random variables.

Let X_1 and X_2 be independent and normally distributed variables, the parameter values being m_1, σ_1 and m_2, σ_2 respectively. Then the sum $X_1 + X_2$ has the d.f. (cf. v, § 1)

$$\Phi\left(\frac{x-m_1}{\sigma_1}\right) * \Phi\left(\frac{x-m_2}{\sigma_2}\right),$$

while the corresponding c.f. is

$$e^{m_1 it - \frac{1}{2}\sigma_1^2 t^2} \cdot e^{m_2 it - \frac{1}{2}\sigma_2^2 t^2} = e^{(m_1+m_2)it - \frac{1}{2}(\sigma_1^2 + \sigma_2^2)t^2}.$$

This is, however, obviously the c.f. of a normal distribution, and so we have the following theorem.

Theorem 17.[1] *The sum of two independent and normally distributed variables is itself normally distributed; thus*

$$\Phi\left(\frac{x-m_1}{\sigma_1}\right) * \Phi\left(\frac{x-m_2}{\sigma_2}\right) = \Phi\left(\frac{x-m}{\sigma}\right),$$

where

$$m = m_1 + m_2, \quad \sigma^2 = \sigma_1^2 + \sigma_2^2.$$

[1] This theorem is sometimes attributed to d'Ocagne, but it seems to have been known already to Poisson and Cauchy, and possibly also to Gauss.

Obviously this theorem is immediately generalized to the composition of any finite number of normal distributions.

We shall now prove three theorems which attach themselves in a natural way to Theorem 17 and reveal further remarkable properties of the normal distribution.

According to Theorem 17, the d.f.'s of the type $\Phi\left(\dfrac{x-m}{\sigma}\right)$ form a *closed family* (the "normal family") with respect to the operation of convolution. Now, any d.f. with a finite mean value m and a finite s.d. σ may be written in the form $F\left(\dfrac{x-m}{\sigma}\right)$, where $F(x)$ is a d.f. with the mean value 0 and the s.d. 1. For any given $F(x)$ with these properties, all functions $F\left(\dfrac{x-m}{\sigma}\right)$ may be considered as a *family* generated by $F(x)$. Our next three theorems then assert (1) that no $F(x)$ different from $\Phi(x)$ generates in this way a *closed* family; (2) that the convolution of any two d.f.'s which do not both belong to the normal family never produces a member of that family; and (3) that every d.f. with a finite s.d. gives, by n-fold convolution with itself, a d.f. which for all sufficiently large n comes (uniformly for all real x) as close as we please to a member of the normal family. We shall first give the formal statements of the three theorems and then proceed to the proofs.

Theorem 18.[1] *Let $F(x)$ be a d.f. with the mean value 0 and the s.d. 1. If, to any constants m_1, m_2 (real) and σ_1, σ_2 (positive), we can find m and σ such that*

$$(53) \qquad F\left(\frac{x-m_1}{\sigma_1}\right) * F\left(\frac{x-m_2}{\sigma_2}\right) = F\left(\frac{x-m}{\sigma}\right),$$

then $F(x) \equiv \Phi(x)$.

---

[1] Pólya [1]. The example of Cauchy's distribution (v, § 6) shows that, in this theorem, it is essential that we consider only d.f.'s with finite s.d.'s Further examples of non-normal d.f.'s satisfying (53) (known as stable distributions) will be found e.g. in Feller's book quoted on p. 118.

Theorem 19.[1] *If the sum of two independent random variables is normally distributed, then each variable is itself normally distributed. Thus if $F_1(x)$ and $F_2(x)$ are d.f.'s such that*

$$(54) \qquad F_1(x) * F_2(x) = \Phi\left(\frac{x-m}{\sigma}\right),$$

then $F_1(x) = \Phi\left(\dfrac{x-m_1}{\sigma_1}\right)$, $F_2(x) = \Phi\left(\dfrac{x-m_2}{\sigma_2}\right)$, *where* $m_1 + m_2 = m$, $\sigma_1^2 + \sigma_2^2 = \sigma^2$.

Before stating the third theorem, some preliminary remarks are necessary. Denoting the composition $F * F * \ldots * F$ of n equal components by F^{n*}, we obtain from Theorem 17

$$\left(\Phi\left(\frac{x-m}{\sigma}\right)\right)^{n*} = \Phi\left(\frac{x-mn}{\sigma\sqrt{n}}\right),$$

and in particular for $m = 0$, $\sigma = 1/\sqrt{n}$,

$$(55) \qquad (\Phi(x\sqrt{n}))^{n*} = \Phi(x).$$

The last relation expresses that if $X_1, \ldots, X_n$ are independent variables, all with the same d.f. $\Phi(x)$, then the variable

$$(X_1 + \ldots + X_n)/\sqrt{n}$$

has the d.f. $\Phi(x)$.

Theorem 20.[2] *Let $F(x)$ be a d.f. with the mean value 0 and the s.d. 1. If X_1, X_2, ... are independent variables all having the d.f. $F(x)$, then the d.f. of the variable $(X_1 + \ldots + X_n)/\sqrt{n}$ tends to $\Phi(x)$ as $n \to \infty$, uniformly for all real x. Thus*

$$(55a) \qquad (F(x\sqrt{n}))^{n*} \to \Phi(x)$$

uniformly in x. Hence it follows also that

$$(56) \qquad \left(F\left(\frac{x-m}{\sigma}\right)\right)^{n*} - \Phi\left(\frac{x-mn}{\sigma\sqrt{n}}\right) \to 0$$

uniformly in x, for all fixed m and σ.

---

[1] Cramér [5]. The theorem had been conjectured by Lévy [2], [3]. It will be observed that in this theorem it is not assumed that the moments of any order are finite.

[2] Lindeberg [1], Lévy [1], p. 233.

Theorem 20 is a particular case of the famous "Central Limit Theorem" in the theory of probability, which will be more fully treated in the following paragraph. We shall now first prove Theorem 20, which will then be used for the proof of Theorem 18. Finally, we shall prove Theorem 19.

Proof of Theorem 20. If $f(t)$ is the c.f. of a d.f. $F(x)$ with $\alpha_1 = 0$ and $\alpha_2 = 1$, it follows from formula (25) of IV, § 2, that $f(t) = 1 - \frac{1}{2}t^2 + o(t^2)$ for small values of $|t|$. Thus we have uniformly in every finite t-interval

$$f\left(\frac{t}{\sqrt{n}}\right) = 1 - \frac{t^2}{2n} + o\left(\frac{1}{n}\right)$$

as $n \to \infty$. The c.f. of the variable $(X_1 + \ldots + X_n)/\sqrt{n}$ is

$$\left(f\left(\frac{t}{\sqrt{n}}\right)\right)^n = \left(1 - \frac{t^2 + o(1)}{2n}\right)^n.$$

As $n \to \infty$, this tends uniformly in every finite t-interval to the limit $e^{-\frac{t^2}{2}}$, which is the c.f. of $\Phi(x)$. Thus by Theorem 11 the d.f. of $(X_1 + \ldots + X_n)/\sqrt{n}$ tends to $\Phi(x)$. The uniformity of the convergence follows easily from the fact that $\Phi(x)$ is continuous. Thus (55a) is proved, and (56) follows immediately from the remark that $\left(F\left(\frac{x-m}{\sigma}\right)\right)^{n*}$ is the d.f. of the variable

$$mn + \sigma\sqrt{n}\,\frac{X_1 + \ldots + X_n}{\sqrt{n}}.$$

Proof of Theorem 18. Both members of the relation (53) are d.f.'s, and the first order moments are $m_1 + m_2$ and m respectively, while the variances are $\sigma_1^2 + \sigma_2^2$ and σ^2, so that we obtain $m = m_1 + m_2$, $\sigma^2 = \sigma_1^2 + \sigma_2^2$. Putting $m_1 = m_2 = \ldots = 0$, we obtain by iteration

$$F\left(\frac{x}{\sigma_1}\right) * \ldots * F\left(\frac{x}{\sigma_n}\right) = F\left(\frac{x}{\sqrt{\sigma_1^2 + \ldots + \sigma_n^2}}\right),$$

and thus in particular

$$(F(x\sqrt{n}))^{n*} = F(x).$$

From (55a) it then follows that $F(x) = \Phi(x)$ for all x.

Proof of Theorem 19. Let X_1 and X_2 be two independent variables with the d.f.'s F_1 and F_2, and the c.f.'s f_1 and f_2, and suppose that $X_1 + X_2$ has the d.f. $\Phi\left(\dfrac{x-m}{\sigma}\right)$. Since the quadrant $X_1 \leqq x$, $X_2 \leqq y$ is a sub-set of the half-plane $X_1 + X_2 \leqq x+y$, we have for all values of x and y

$$F_1(x) F_2(y) \leqq \Phi\left(\frac{x+y-m}{\sigma}\right).$$

Here we choose for y any fixed value such that $F_2(y) > 0$, and use the inequality

$$\Phi(x) < \frac{1}{\sqrt{2\pi}\,|x|} e^{-\frac{x^2}{2}},$$

which holds for all $x < 0$ and is easily proved by partial integration. It then follows that we can determine A and B independent of x, such that for all $x < 0$

$$F_1(x) < A e^{-\frac{x^2}{2\sigma^2} + B|x|}.$$

Similarly we can determine A' and B' such that for all $x > 0$

$$1 - F_1(x) < A' e^{-\frac{x^2}{2\sigma^2} + B'x}.$$

From the two last inequalities it follows that the integral

(57) $$J = \int_{-\infty}^{\infty} e^{\frac{x^2}{4\sigma^2}} dF_1(x)$$

is convergent. If, now, we consider the c.f.

$$f_1(t) = \int_{-\infty}^{\infty} e^{itx} dF_1(x)$$

for *complex values of the variable t*, it follows from the convergence of (57) that the integral which represents $f_1(t)$ is absolutely and uniformly convergent in every finite domain in the t-plane. Thus

E

$f_1(t)$ is an *integral function* of the complex variable t. For the modulus of this function we obtain by means of the elementary inequality

$$|tx| \leq \sigma^2 |t|^2 + \frac{x^2}{4\sigma^2},$$

$$|f_1(t)| \leq \int_{-\infty}^{\infty} e^{\sigma^2|t|^2 + \frac{x^2}{4\sigma^2}} dF_1(x) = J e^{\sigma^2|t|^2},$$

so that *the order*[1] *of the integral function $f_1(t)$ does not exceed* 2. In the same way it is proved that $f_2(t)$ is an integral function of t, the order of which does not exceed 2. According to (54) we have, however,

$$f_1(t) f_2(t) = e^{mit - \frac{1}{2}\sigma^2 t^2},$$

which shows that f_1 and f_2 are *integral functions without zeros.* By the classical factorization theorem[2] of Hadamard it then follows that

(58) $$f_1(t) = e^{q_1(t)}, \quad f_2(t) = e^{q_2(t)},$$

where $q_1(t)$ and $q_2(t)$ are polynomials of degree not greater than 2.

The convergence of (57) implies that all moments and semi-invariants of X_1 are finite. Denoting the mean value by m_1 and the s.d. by σ_1, we then obtain from (58) according to IV, § 2,

IV, § 2, $$f_1(t) = e^{m_1 it - \frac{1}{2}\sigma_1^2 t^2},$$

and similarly $$f_2(t) = e^{m_2 it - \frac{1}{2}\sigma_2^2 t^2}.$$

This is, however, equivalent to

$$F_1(x) = \Phi\left(\frac{x - m_1}{\sigma_1}\right), \quad F_2(x) = \Phi\left(\frac{x - m_2}{\sigma_2}\right).$$

Then obviously $m_1 + m_2 = m$ and $\sigma_1^2 + \sigma_2^2 = \sigma^2$, and the theorem is proved.[3]

---

[1] Cf. e.g. Titchmarsh [1], p. 248.   [2] Cf. e.g. Titchmarsh [1], p. 250.

[3] σ_1 or σ_2 may be equal to zero. If, e.g., $\sigma_1 = 0$, we have by § 1 to interpret $\Phi\left(\frac{x - m_1}{\sigma_1}\right)$ as $\epsilon(x - m_1)$, and so obtain the trivial solution of (54): $F_1(x) = \epsilon(x - m_1)$, $F_2(x) = \Phi\left(\frac{x - m + m_1}{\sigma}\right)$.

3. The **Central Limit Theorem**[1] in the theory of probability asserts that, under certain general conditions, *the sum of a large number of independent variables is approximately normally distributed*. In Theorem 20, we have already met with a particular case of the general theorem, viz. the convolution of n equal components with a finite s.d. We shall now consider the case when the components are not necessarily equal. *Throughout this paragraph and the immediately following one, we shall suppose that every component has a finite s.d. and a mean value equal to zero.* The assumption that the mean value is zero may obviously be made without loss of generality, since it is equivalent to the simple addition of a constant to each variable.

We thus consider a sequence of independent random variables X_1, X_2, ..., such that X_ν has the mean value 0 and the s.d. σ_ν. The d.f. of X_ν will be denoted by $F_\nu(x)$ and the c.f. by $f_\nu(t)$.

If the d.f. of the sum $X_1 + \ldots + X_n$ is denoted by $\overline{F}_n(x)$, we have

(59) $$\overline{F}_n(x) = F_1(x) * F_2(x) * \ldots * F_n(x),$$

and $\overline{F}_n(x)$ has the mean value zero and the variance s_n^2 given by

(60) $$s_n^2 = \sigma_1^2 + \sigma_2^2 + \ldots + \sigma_n^2.$$

The variable $(X_1 + \ldots + X_n)/s_n$ then has the d.f.

(61) $$\mathfrak{F}_n(x) = \overline{F}_n(s_n x)$$

with the mean value 0 and the s.d. 1. *It is possible to show that under fairly general conditions $\mathfrak{F}_n(x)$ tends to the normal d.f.* $\Phi(x)$ *as n tends to infinity.* The most important case is that in which the following two conditions are satisfied:

(62) $$s_n \to \infty, \quad \text{and} \quad \frac{\sigma_n}{s_n} \to 0.$$

---

[1] This theorem was first stated by Laplace, and was further treated by several mathematicians during the nineteenth century, notably Tchebycheff and Markoff. A complete and rigorous proof under fairly general conditions was first given in 1901 by Liapounoff [1], [2]. Cf. VI, § 4, and VII, § 4. A comprehensive account of the modern development of the subject is given, for example, in the books by Gnedenko-Kolmogoroff, Loève and Chung quoted on p. 118.

It is easily shown that the two conditions (62) are equivalent to the single condition

(62 a) $\dfrac{\sigma_\nu}{s_n} \to 0$ uniformly for $\nu = 1, 2, \ldots, n$.

This means that the total s.d. of $\overset{n}{\underset{1}{\Sigma}} X_\nu$ tends to infinity, while each component contributes only a small fraction of the total s.d. In this case, we have the following theorem.

Theorem 21.[1] *Let X_1, X_2, ... be a sequence of independent random variables with vanishing mean values and finite s.d.'s satisfying (62), and denote by $\mathfrak{F}_n(x)$ the d.f. of the variable $(X_1 + \ldots + X_n)/s_n$ as defined by (59) and (61). Then a necessary and sufficient condition for the validity of the relation*

(63) $\lim\limits_{n \to \infty} \mathfrak{F}_n(x) = \Phi(x)$

for all x is that, for any given $\epsilon > 0$,

(64) $\lim\limits_{n \to \infty} \dfrac{1}{s_n^2} \overset{n}{\underset{1}{\Sigma}} \int_{|x| > \epsilon s_n} x^2 \, dF_\nu(x) = 0.$

It is readily seen that Theorem 20 is contained as a particular case in Theorem 21. The condition (64) is known as the *Lindeberg condition*. It is here given in a slightly simpler form than that originally given by Lindeberg.

In order to prove the theorem, we denote by $\mathfrak{f}_n(t)$ the c.f. which corresponds to the d.f. $\mathfrak{F}_n(t)$, and then obtain from (59) and (61)

(65) $\mathfrak{f}_n(t) = f_1(t/s_n) f_2(t/s_n) \ldots f_n(t/s_n).$

Now, for any integer $k > 0$ and for any real a we have

(66) $e^{ia} = \overset{k-1}{\underset{0}{\Sigma}} \dfrac{(ia)^\nu}{\nu!} + \vartheta \dfrac{a^k}{k!},$

using ϑ as a general notation for a real or complex quantity of modulus not exceeding unity. We shall first prove that the condition is *sufficient*, and thus assume that (64) is satisfied for any given $\epsilon > 0$. Taking in (66) $k = 2$ for $|x| > \epsilon s_n$ and $k = 3$ for $|x| \leq \epsilon s_n$,

[1] Lindeberg [1], Lévy [1], Feller [1]. It can be shown without difficulty that the condition (64) implies (62). Thus (64) is necessary and sufficient for the joint validity of (62) and (63).

we obtain for $|t| < T$, where $T > 1$,

$$f_\nu(t/s_n) = \int_{-\infty}^{\infty} e^{\frac{itx}{s_n}} dF_\nu$$

$$= 1 - \frac{t^2}{2s_n^2} \int_{|x| \le \epsilon s_n} x^2 dF_\nu + \vartheta \frac{T^3}{6s_n^3} \int_{|x| \le \epsilon s_n} |x|^3 dF_\nu$$

$$+ \vartheta \frac{T^2}{2s_n^2} \int_{|x| > \epsilon s_n} x^2 dF_\nu$$

$$= 1 - \frac{\sigma_\nu^2}{2s_n^2} t^2 + \frac{\vartheta T^3}{s_n^2} \left(\epsilon \sigma_\nu^2 + \int_{|x| > \epsilon s_n} x^2 dF_\nu \right),$$

bearing in mind that the mean value of X_ν is equal to zero. By (62a) we then conclude that $f_\nu(t/s_n) \to 1$, uniformly for $|t| < T$ and $\nu = 1, 2, \ldots, n$. It follows that we have

$$\log f_\nu(t/s_n) = (1 + \eta)(f_\nu(t/s_n) - 1),$$

where $|\eta| < \epsilon$ for all sufficiently large n. As we may obviously suppose $0 < \epsilon < \frac{1}{2}$, we thus obtain

$$\log f_\nu(t/s_n) = -\frac{\sigma_\nu^2}{2s_n^2} t^2 + \frac{2\vartheta T^3}{s_n^2} \left(\epsilon \sigma_\nu^2 + \int_{|x| > \epsilon s_n} x^2 dF_\nu \right).$$

Summing over $\nu = 1, 2, \ldots, n$, we obtain according to (65) for $0 < \epsilon < \frac{1}{2}$ and $|t| < T$

$$\log \mathfrak{f}_n(t) = -\frac{t^2}{2} + 2\vartheta T^3 \left(\epsilon + \frac{1}{s_n^2} \sum_1^n \int_{|x| > \epsilon s_n} x^2 dF_\nu \right).$$

ϵ being arbitrary, it then follows from (64) that we have as $n \to \infty$

$$(67) \qquad \log \mathfrak{f}_n(t) = \sum_1^n \log f_\nu(t/s_n) \to -\frac{t^2}{2}$$

uniformly for $|t| < T$, and by Theorem 11 this is equivalent to (63). Thus the condition (64) is *sufficient*.

In order to prove that (64) is also *necessary*, we assume that (63) and thus also (67) is satisfied. Using (66) with $k = 2$, we have

$$f_\nu(t/s_n) = 1 - \vartheta \frac{\sigma_\nu^2 t^2}{2s_n^2},$$

so that $f_\nu(t/s_n) \to 1$, uniformly in the same sense as above, while $\sum_1^n |f_\nu(t/s_n) - 1|$ is bounded for every fixed t. Since $(z-1)/\log z \to 1$ as $z \to 1$, it then follows from (67) that

$$(68) \qquad \sum_1^n (f_\nu(t/s_n) - 1) \to -\frac{t^2}{2}$$

for every real t. According to the Bienaymé-Tchebycheff inequality (III, § 3) we have, however, paying regard to (60),

$$\sum_1^n \int_{|x| > \epsilon s_n} dF_\nu(x) \leq \frac{1}{\epsilon^2},$$

and so obtain from (68), taking the real part,

$$(69) \qquad \limsup_{n \to \infty} \left| \frac{t^2}{2} - \sum_1^n \int_{|x| \leq \epsilon s_n} \left(1 - \cos\frac{tx}{s_n}\right) dF_\nu \right| \leq \frac{2}{\epsilon^2}.$$

On the other hand, we have

$$\sum_1^n \int_{|x| \leq \epsilon s_n} \left(1 - \cos\frac{tx}{s_n}\right) dF_\nu \leq \frac{t^2}{2s_n^2} \sum_1^n \int_{|x| \leq \epsilon s_n} x^2 dF_\nu \leq \frac{t^2}{2}.$$

Introducing this in (69), we obtain

$$0 \leq \frac{t^2}{2} \limsup_{n \to \infty} \frac{1}{s_n^2} \sum_1^n \int_{|x| > \epsilon s_n} x^2 dF_\nu \leq \frac{2}{\epsilon^2}.$$

Since t may be taken arbitrarily large, it follows that the condition (64) must be satisfied, and thus Theorem 21 is proved.

If the conditions (62) are not satisfied, one of the following two cases must occur:

(A) $\quad \lim_{n \to \infty} s_n = s$ exists; or

(B) $\quad s_n \to \infty$, $\sigma_n/s_n > \alpha > 0$ for an infinity of values of n.

In the case (A), it follows from (61) that the relation $\mathfrak{F}_n(x) \to \Phi(x)$ is equivalent to $\overline{F}_n(x) \to \Phi(x/s)$. Thus by (59) the infinite convolution $F_1(x) * F_2(x) * \ldots$ converges (V, § 4) to $\Phi(x/s)$. Putting $G(x) = F_2(x) * F_3(x) * \ldots$, we then have $F_1(x) * G(x) = \Phi(x/s)$. From Theorem 19 we then obtain $F_1(x) = \Phi(x/\sigma_1)$, and it follows that *in case (A) the necessary and sufficient condition for $\mathfrak{F}_n(x) \to \Phi(x)$ is that each variable X_ν is normally distributed.*

In case (B), on the other hand, for values of n such that $\sigma_n/s_n > \alpha > 0$, the s.d. of X_n is not small compared to the total s.d. of $\sum_1^n X_\nu$. It is then easily understood that (63) cannot be satisfied unless the d.f.'s of these "large" X_n tend to the normal type. We shall, however, not enter upon a detailed discussion of this case.

4. A *sufficient* condition for the validity of (63), which is often useful, has been given by Liapounoff.[1] Let $\beta_{k\nu} = E\,(|\,X_\nu\,|^k)$ denote the absolute moment of order k of the d.f. $F_\nu\,(x)$, so that in particular $\beta_{2\nu} = \sigma_\nu^2$. Suppose that for some $k > 2$ (not necessarily integral) $\beta_{k\nu}$ is finite for all ν and is such that

$$(70) \qquad \frac{1}{s_n^k} \sum_{\nu=1}^n \beta_{k\nu} = \frac{\sum\limits_1^n \beta_{k\nu}}{\left(\sum\limits_1^n \beta_{2\nu}\right)^{k/2}} \to 0$$

as $n \to \infty$. We then have

$$\frac{1}{s_n^2} \sum_1^n \int_{|x| > \epsilon s_n} x^2 dF_\nu \leqq \frac{1}{\epsilon^{k-2} s_n^k} \sum_1^n \int_{|x| > \epsilon s_n} |\,x\,|^k\, dF_\nu \leqq \frac{1}{\epsilon^{k-2} s_n^k} \sum_1^n \beta_{k\nu},$$

and thus the Lindeberg condition (64) is satisfied, so that by Theorem 21 we have $\mathfrak{F}_n\,(x) \to \Phi\,(x)$.

If, in particular, there are two positive constants M and m, such that for all ν we have $\beta_{k\nu} < M$ and $\beta_{2\nu} = \sigma_\nu^2 > m$, it is obvious that the Liapounoff condition (70) is satisfied, and thus $\mathfrak{F}_n\,(x)$ tends to $\Phi\,(x)$.

5. We shall now apply the results of the two preceding paragraphs to some particular examples.

As a first example, we take the variables $X_r = Y_r - p_r$, where Y_r has the simple discontinuous distribution considered in v, § 5: $Y_r = 1$ with the probability p_r, and $Y_r = 0$ with the probability

---

[1] Liapounoff [2]. By means of the condition (70), Liapounoff obtained an upper limit for the modulus of the difference $\mathfrak{F}_n\,(x) - \Phi\,(x)$. This question will be considered in the following Chapter (cf. Theorem 24).

$q_r = 1 - p_r$. We then have $E(X_r) = 0$, $D^2(X_r) = \sigma_r^2 = p_r q_r$ and $\beta_{3r} = E(|X_r|^3) = p_r q_r (p_r^2 + q_r^2) \leqq p_r q_r$, so that

$$\frac{1}{s_n^3} \sum_1^n \beta_{3r} \leqq \left(\sum_1^n p_r q_r \right)^{-\frac{1}{2}}$$

Putting, as in v, §5, $\nu = Y_1 + \ldots + Y_n$, so that ν represents the number of those Y_r which assume the value 1, we have

$$U_n = \frac{X_1 + \ldots + X_n}{s_n} = \frac{\nu - \sum_1^n p_r}{\left(\sum_1^n p_r q_r \right)^{\frac{1}{2}}}.$$

If the series $\sum p_r q_r$ is *divergent*, the Liapounoff condition (70) is satisfied, and thus the d.f. of the variable U_n tends to $\Phi(x)$ as $n \to \infty$. If, on the other hand, $\sum p_r q_r$ is *convergent*, it follows from the above discussion (case (A), p. 59) that the d.f. of U_n does *not* tend to $\Phi(x)$, since the variables X_r are not normally distributed.

In the particular case when all p_r are equal to p, where $0 < p < 1$, the series $\sum p_r q_r$ is obviously divergent, so that the d.f. of the variable $U_n = (\nu - np)/\sqrt{npq}$ tends to $\Phi(x)$. It follows that for any fixed λ_1 and λ_2 the probability of the relation

$$\lambda_1 < (\nu - np)/\sqrt{npq} < \lambda_2$$

tends to the limit $\dfrac{1}{\sqrt{2\pi}} \displaystyle\int_{\lambda_1}^{\lambda_2} e^{-\frac{t^2}{2}} dt$. This is the extended form of Bernoulli's theorem proved by De Moivre and Laplace.

As a second example we consider the variables X_r with the distribution

$$X_r = \begin{cases} -r^\alpha \text{ with the probability } \dfrac{1}{2r^{2\alpha}}, \\ 0 \qquad\qquad ,, \qquad\qquad 1 - \dfrac{1}{r^{2\alpha}}, \\ r^\alpha \qquad\qquad ,, \qquad\qquad \dfrac{1}{2r^{2\alpha}}. \end{cases}$$

Obviously $E(X_r) = 0$ and $D^2(X_r) = \sigma_r^2 = 1$, so that

$$s_n^2 = \sigma_1^2 + \ldots + \sigma_n^2 = n.$$

Thus (62) is satisfied, and by Theorem 21 a necessary and sufficient condition for $\mathfrak{F}_n(x) \to \Phi(x)$ is

$$\lim_{n \to \infty} \frac{1}{n} \sum_{\substack{1 \le \nu \le n \\ \nu^\alpha > \epsilon \sqrt{n}}} 1 = 0.$$

It is readily seen that this condition is satisfied for $\alpha < \frac{1}{2}$, but not
for $\alpha \ge \frac{1}{2}$. For $\alpha > \frac{1}{2}$, it is indeed obvious that the distribution
cannot tend to the normal type, as in this case we have a prob-
ability greater than $\prod_2^\infty (1 - r^{-2\alpha}) > 0$ that any sum $X_2 + \ldots + X_n$
assumes the value zero. The Liapounoff condition (70) is satisfied
for $\alpha < \frac{1}{2}$, but not for $\alpha \ge \frac{1}{2}$.

6. If, for the independent variables X_n considered in Theorem
21, the existence of finite mean values and variances is not
assumed, we may still ask if it is possible to find constants a_n and
b_n such that the d.f. of $(X_1 + \ldots + X_n)/a_n - b_n$ tends to $\Phi(x)$ as
$n \to \infty$. The same question may, of course, be asked in a case
when finite mean values and variances do exist, but the Lindeberg condition (64) is not satisfied. We shall not enter upon a
detailed discussion of the problems belonging to this order of
ideas,.but shall content ourselves with proving the following two
theorems.

Theorem 22.[1] *Let X_1, X_2, ... be a sequence of independent
random variables, and denote by $F_\nu(x)$ the d.f. of X_ν. If, for a
sequence $a_1, a_2, \ldots$ of positive numbers, the conditions*

$$(71) \qquad \lim_{n \to \infty} \sum_{\nu=1}^n \int_{|x| > \epsilon a_n} dF_\nu(x) = 0,$$

$$(72) \qquad \lim_{n \to \infty} \frac{1}{a_n^2} \sum_{\nu=1}^n \int_{|x| \le \epsilon a_n} x^2 dF_\nu(x) = 1,$$

$$(73) \qquad \lim_{n \to \infty} \frac{1}{a_n} \sum_{\nu=1}^n \left| \int_{|x| \le \epsilon a_n} x dF_\nu(x) \right| = 0,$$

---

[1] Feller [1]. It is there further shown that (71)–(73) are *necessary* for the con-
vergence to $\Phi(x)$ of the d.f. of any variable $(\delta_1 X_1 + \ldots + \delta_n X_n)/a_n$, where $\delta_i = \pm 1$.

are satisfied for every $\epsilon > 0$, then the d.f. of the variable

$$(X_1 + \ldots + X_n)/a_n$$

tends to $\Phi(x)$ as $n \to \infty$.

If, in the particular case when every X_ν has a finite s.d. and a mean value equal to zero, we take $a_n = s_n$ as in Theorem 21, it is easily found that the conditions (71)–(73) reduce to the Lindeberg condition (64).

In order to prove the theorem, we denote by $f_\nu(t)$ the c.f. of X_ν. According to (71) we may, to any given $\epsilon > 0$, choose $n_0 = n_0(\epsilon)$ such that for all $n > n_0$

$$\sum_1^n \int_{|x| > \epsilon a_n} dF_\nu < \epsilon.$$

For any fixed t and for $\nu = 1, 2, \ldots, n$ we then obtain, by means of (66) and (71)–(73), the following three inequalities:

$$|f_\nu(t/a_n) - 1| \leq \int_{-\infty}^{\infty} |e^{\frac{itx}{a_n}} - 1| \, dF_\nu$$

$$= \int_{|x| > \epsilon a_n} + \int_{|x| \leq \epsilon a_n} < (2 + |t|)\epsilon,$$

$$\sum_1^n |f_\nu(t/a_n) - 1| < 2\epsilon + \sum_1^n \left| \int_{|x| \leq \epsilon a_n} (e^{\frac{itx}{a_n}} - 1) \, dF_\nu \right|$$

$$= 2\epsilon + \sum_1^n \left| \int_{|x| \leq \epsilon a_n} \left(\frac{itx}{a_n} + \vartheta \frac{t^2 x^2}{2a_n^2} \right) dF_\nu \right|$$

$$\leq 2\epsilon + \frac{|t|}{a_n} \sum_1^n \left| \int_{|x| \leq \epsilon a_n} x \, dF_\nu \right| + \frac{t^2}{2a_n^2} \sum_1^n \int_{|x| \leq \epsilon a_n} x^2 \, dF_\nu < (2 + |t|)\epsilon + t^2,$$

$$\left| \sum_1^n (f_\nu(t/a_n) - 1) + \frac{t^2}{2} \right|$$

$$< 2\epsilon + \left| \sum_1^n \int_{|x| \leq \epsilon a_n} (e^{\frac{itx}{a_n}} - 1) \, dF_\nu + \frac{t^2}{2} \right|$$

$$\leq 2\epsilon + \frac{|t|}{a_n} \sum_1^n \left| \int_{|x| \leq \epsilon a_n} x \, dF_\nu \right| + \frac{t^2}{2} \left| \frac{1}{a_n^2} \sum_1^n \int_{|x| \leq \epsilon a_n} x^2 \, dF_\nu - 1 \right|$$

$$< (2 + |t| + t^2)\epsilon + \frac{|t|^3 \epsilon}{6a_n^2} \sum_1^n \int_{|x| \leq \epsilon a_n} x^2 \, dF_\nu < (2 + |t| + t^2 + |t|^3)\epsilon,$$

for all sufficiently large n. Thus we have, as $n \to \infty$,

$$f_\nu(t/a_n) - 1 \to 0,$$

$$\sum_{\nu=1}^{n}(f_\nu(t/a_n) - 1) \to -\frac{t^2}{2},$$

and $$\limsup \sum_{\nu=1}^{n}|f_\nu(t/a_n) - 1| \leq t^2,$$

uniformly for $\nu = 1, 2, ..., n$. It follows that for every t

$$\sum_{\nu=1}^{n}\log f_\nu(t/a_n) \to -\frac{t^2}{2},$$

or $$\prod_{\nu=1}^{n}f_\nu(t/a_n) \to e^{-\frac{t^2}{2}},$$

The first member of the last relation is, however, the c.f. of the variable $(X_1 + ... + X_n)/a_n$, and thus by Theorem 11 the theorem is proved.

We shall now consider the case when all the variables X_ν have the same probability distribution.

Theorem 23.[1] *Let* X_1, X_2, ... *be a sequence of independent variables all having the same d.f.* $F(x)$. *If we have*

$$(74) \qquad \int_{|x|>z}dF(x) \rightleftharpoons o\left(\frac{1}{z^2}\int_{|x|\leq z}x^2 dF(x)\right)$$

as $z \to \infty$, *then:*

(I) *The absolute moment* $\beta_r = \int_{-\infty}^{\infty}|x|^r dF(x)$ *is finite for* $0 \leq r < 2$,

so that in particular a finite mean value $m = \int_{-\infty}^{\infty}x dF(x)$ *exists.*

(II) *It is possible to find a sequence* $a_1, a_2, ...$ *of positive numbers such that the d.f. of the variable*

$$(75) \qquad U_n = \frac{X_1 + ... + X_n - mn}{a_n}$$

tends to $\Phi(x)$ *as* $n \to \infty$.

[1] Feller [1], [2], Khintchine [3], Lévy [3]. It is shown by these authors that (74) is also a *necessary* condition for·the existence of two sequences $\{a_n\}$ and $\{b_n\}$ such that the d.f. of $(X_1 + ... + X_n)/a_n - b_n$ tends to $\Phi(x)$. On the other hand, (74) is *not* a necessary condition for the convergence of β_r for $0 \leq r < 2$.

For the proof of this theorem, we may obviously assume that β_2 is *not* finite, as otherwise (I) is trivial and (II) is an immediate corollary of Theorem 20.

We shall first prove that β_r is finite for $0 \leqq r < 2$. The function

$$\psi(z) = \int_{|x| \leqq z} x^2 dF(x) = -z^2 \int_{|x| > z} dF(x) + 2 \int_0^z v \, dv \int_{|x| > v} dF(x)$$

is never decreasing for $z > 0$ and tends to infinity with z. By (74) we have $(z \to \infty)$

$$\psi(z) \leqq 2 \int_0^z v \, dv \int_{|x| > v} dF(x) = o \left(\int_1^z \frac{\psi(v)}{v} dv \right).$$

$\epsilon > 0$ being given, we denote by $M(z)$ the upper bound of $v^{-\epsilon} \psi(v)$ in the interval $1 \leqq v \leqq z$, and then obtain

$$\int_1^z \frac{\psi(v)}{v} dv \leqq M(z) \int_1^z v^{\epsilon-1} dv < \frac{z^\epsilon M(z)}{\epsilon}.$$

Thus we have $z^{-\epsilon} \psi(z) = o(M(z))$, which shows that $\psi(z) = o(z^\epsilon)$ for every $\epsilon > 0$. It follows that, for any fixed r such that $0 \leqq r < 2$ and for all sufficiently large z,

$$\int_{z < |x| \leqq 2z} |x|^r dF(x) < z^{r-2} \psi(2z) < z^{r/2-1},$$

and this obviously implies that β_r is finite. Hence in particular the mean value m is finite.

We now proceed to prove the assertion (II). As by hypothesis β_2 is not finite, the first member of (74) is positive for all $z > 0$, and the function

(76)
$$Z(u) = \text{lower bound of all } z > 0 \text{ such that } \int_{|x| > z} dF(x) \leqq u,$$

is a positive and never increasing function of u, uniquely defined for $0 < u < 1$ and tending to infinity as u tends to zero. Further, according to (74) we can find a steadily decreasing function

$\eta\,(z)$, tending to zero as $z \to \infty$, such that for all $z > 0$

$$(77) \qquad \int_{|x|>z} dF\,(x) < \frac{\eta\,(z)}{z^2} \int_{|x|\leq z} x^2 dF\,(x).$$

Let $\{\lambda_n\}$ denote a decreasing sequence of numbers such that $0 < \lambda_n < 1$ and

$$(78) \qquad \lambda_n \to 0, \quad \frac{\lambda_n}{\eta\left(\tfrac{1}{2} Z\left(\dfrac{1}{n}\right)\right)} \to \infty.$$

We put

$$(79) \qquad z_n = Z\,(\lambda_n/n), \quad a_n^2 = n \int_{|x|\leq z_n} x^2 dF\,(x),$$

and are now going to show that, with this definition of a_n, the d.f. of the variable U_n defined by (75) tends to $\Phi\,(x)$. Putting $\bar{X}_\nu = X_\nu - m$, we have $U_n = (\bar{X}_1 + \ldots + \bar{X}_n)/a_n$, the d.f. of each $\bar{X}_\nu$ being $F\,(x+m)$. We now apply Theorem 22 to the sequence $\bar{X}_1, \bar{X}_2, \ldots$, and then only have to show that the conditions (71)–(73) are satisfied if we put $F_\nu\,(x) = F\,(x+m)$ and define a_n according to (79).

By means of (76) and (79) we obtain

$$(80) \qquad \int_{|x|>z_n} dF\,(x) \leq \frac{\lambda_n}{n}, \quad \int_{|x|>\frac{1}{2}z_n} dF\,(x) > \frac{\lambda_n}{n},$$

and further according to (77)–(79)

$$\frac{a_n^2}{z_n^2} \geq \frac{n}{z_n^2} \int_{|x|\leq\frac{1}{2}z_n} x^2 dF\,(x) > \frac{n}{4\eta\,(\frac{1}{2}z_n)} \int_{|x|>\frac{1}{2}z_n} dF\,(x)$$

$$\geq \frac{\lambda_n}{4\eta\left(\tfrac{1}{2} Z\left(\dfrac{1}{n}\right)\right)} \to \infty,$$

so that $z_n = o\,(a_n)$. $\epsilon > 0$ being given, we now choose n_0 such that for all $n > n_0$ we have $z_n < \tfrac{1}{2}\epsilon a_n$ and $|m| < \tfrac{1}{2}\epsilon a_n$, and then obtain by (80)

$$n \int_{|x|>\epsilon a_n} dF\,(x+m) < n \int_{|x|>z_n} dF\,(x) \to 0,$$

so that (71) is satisfied. We have further for $n > n_0$

$$\left| \frac{n}{a_n^2} \int_{|x| \le \epsilon a_n} x^2 dF(x+m) - 1 \right|$$

$$= \frac{n}{a_n^2} \left| \int_{|x-m| \le \epsilon a_n} (x-m)^2 dF(x) - \int_{|x| \le z_n} x^2 dF(x) \right|$$

$$< \frac{n}{a_n^2} \left| \int_{|x| \le z_n} (m^2 - 2mx) dF(x) \right| + \frac{n}{a_n^2} \epsilon^2 a_n^2 \int_{|x| > z_n} dF(x)$$

$$< (m^2 + 2\beta_1 |m|) \frac{n}{a_n^2} + \epsilon^2 n \int_{|x| > z_n} dF(x).$$

According to (79) and (80) the last expression tends, however, to zero as $n \to \infty$, so that (72) is also satisfied.

Thus it only remains to show that (73) is satisfied. By (74) we have for every fixed $\delta > 0$ and for all sufficiently large z

$$z \int_{|x| > z} |x| dF(x) = z^2 \int_{|x| > z} dF(x) + z \int_z^\infty dv \int_{|x| > v} dF(x)$$

$$< \delta \psi(z) + \delta z \int_z^\infty \frac{\psi(v)}{v^2} dv$$

$$= 2\delta \psi(z) + \delta z \int_{|x| > z} |x| dF(x),$$

and consequently, putting $z = \frac{1}{2} \epsilon a_n$,

$$(81) \qquad a_n \int_{|x| > \frac{1}{2} \epsilon a_n} |x| dF(x) = o(\psi(\tfrac{1}{2} \epsilon a_n))$$

for every fixed $\epsilon > 0$, as $n \to \infty$. By (79) and (80) we have, however, for all $n > n_0$,

$$\frac{n}{a_n^2} \psi(\tfrac{1}{2} \epsilon a_n) = 1 + \frac{n}{a_n^2} \int_{z_n < |x| \le \frac{1}{2} \epsilon a_n} x^2 dF(x)$$

$$< 1 + \epsilon^2 n \int_{|x| > z_n} dF(x) = O(1),$$

and thus by (81)

$$\frac{n}{a_n} \int_{|x|>\frac{1}{2}\epsilon a_n} |x| \, dF(x) \to 0.$$

Finally we have, the mean value of each $\bar{X}_\nu$ being equal to zero,

$$\frac{n}{a_n} \left| \int_{|x| \leqq \epsilon a_n} x \, dF(x+m) \right| = \frac{n}{a_n} \left| \int_{|x|>\epsilon a_n} x \, dF(x+m) \right|$$

$$< \frac{2n}{a_n} \int_{|x|>\epsilon a_n} |x+m| \, dF(x+m) \leqq \frac{2n}{a_n} \int_{|x|>\frac{1}{2}\epsilon a_n} |x| \, dF(x) \to 0.$$

Thus (73) is satisfied, and the proof of Theorem 23 is completed.

CHAPTER VII

ERROR ESTIMATION.
ASYMPTOTIC EXPANSIONS

1. In VI, § 3, we have considered a sequence of independent variables $\{X_n\}$ such that X_n has the d.f. $F_n(x)$ with the mean value zero and the s.d. σ_n. As in VI, § 3, we put

$$\text{and} \qquad s_n^2 = \sigma_1^2 + \ldots + \sigma_n^2$$

$$(82) \qquad \mathfrak{F}_n(x) = F_1(s_n x) * \ldots * F_n(s_n x),$$

so that $\mathfrak{F}_n(x)$ is the d.f. of the variable $(X_1 + \ldots + X_n)/s_n$. The corresponding c.f. is then

$$(83) \qquad \mathfrak{f}_n(t) = f_1(t/s_n) \ldots f_n(t/s_n).$$

If the Lindeberg condition (64) is satisfied, it follows from Theorem 21 that $\mathfrak{F}_n(x)$ tends to the normal function $\Phi(x)$ as $n \to \infty$. It is then natural to try to investigate the asymptotic behaviour of the error involved in replacing $\mathfrak{F}_n(x)$ by $\Phi(x)$. In this respect, it might be desired: (I) to find an upper limit for the modulus of the difference $\mathfrak{F}_n(x) - \Phi(x)$, and (II) to obtain some kind of asymptotic expansion of this difference for large values of n.

In the present Chapter, both these questions will be treated. An upper limit for the error $|\mathfrak{F}_n(x) - \Phi(x)|$ was first given by Liapounoff [1], [2]. In Theorem 24 we shall give an improvement of his estimation, which shows that under certain conditions the error is of the order $n^{-\frac{1}{2}}$ as $n \to \infty$. It will then be shown in Theorems 25 and 26 that, under somewhat more restrictive conditions, an asymptotic expansion of $\mathfrak{F}_n(x) - \Phi(x)$ in powers of $n^{-\frac{1}{2}}$ can be obtained. In the last paragraph of the chapter, the relations between this expansion and the expansions in series of Hermite polynomials will be discussed.

2. *Throughout the whole Chapter, we shall consider a sequence* $X_1, X_2, \ldots$ *of independent random variables such that* X_n *has the mean value zero and the s.d.* σ_n. *The trivial case when all the* σ_n *are equal to zero will always be excluded. The* νth *order moment, absolute moment and semi-invariant (cf.* IV, § 2) *of the variable* X_n *will be denoted by* $\alpha_{\nu n}, \beta_{\nu n}$ *and* $\gamma_{\nu n}$ *respectively. Thus in particular*
$$\alpha_{1n} = \gamma_{1n} = 0, \ \alpha_{2n} = \beta_{2n} = \gamma_{2n} = \sigma_n^2.$$

Throughout the whole Chapter it will be assumed that there exists an integer $k \geqq 3$ *such that* β_{kn} *is finite for all* $n = 1, 2, \ldots$ *It then follows that* $\alpha_{\nu n}, \beta_{\nu n}$ *and* $\gamma_{\nu n}$ *are finite for* $\nu = 1, 2, \ldots, k$. *In the particular case when all moments are finite,* k *may be chosen as large as we please.*

We shall use the letters ϑ *and* Θ_k *to denote unspecified quantities such that* $|\vartheta| \leqq 1$, *while* $|\Theta_k|$ *is less than a number depending only on* k.

All the results of this Chapter take a particularly simple form in the case when all the variables X_n have the same d.f. We shall refer to this case as *the case of equal components*, and the common d.f. of the variables X_n will be denoted by $F(x)$. If, in this case, σ denotes the s.d. of X_n, we have $s_n = \sigma \sqrt{n}$, and the relations (82) and (83) become
$$\mathfrak{F}_n(x) = (F(\sigma x \sqrt{n}))^{n*}, \quad \mathfrak{f}_n(t) = (f(t/(\sigma \sqrt{n})))^n.$$

3. In this paragraph we shall deduce some lemmas that are required for the proofs of the results indicated in § 1. We put for $\nu = 2, 3, \ldots, k$

$$(84) \quad \begin{cases} B_{\nu n} = \dfrac{1}{n}(\beta_{\nu 1} + \ldots + \beta_{\nu n}), \quad \Gamma_{\nu n} = \dfrac{1}{n}(\gamma_{\nu 1} + \ldots + \gamma_{\nu n}), \\[2mm] \rho_{\nu n} = \dfrac{B_{\nu n}}{B_{2n}^{\nu/2}}, \qquad\qquad\qquad \lambda_{\nu n} = \dfrac{\Gamma_{\nu n}}{\Gamma_{2n}^{\nu/2}}. \end{cases}$$

Thus for $\nu = 2$ we have $B_{2n} = \Gamma_{2n} = s_n^2/n$, $\rho_{2n} = \lambda_{2n} = 1$. $B_{\nu n}$ is the νth absolute moment of the d.f. $(F_1(x) + \ldots + F_n(x))/n$, and thus by (20) $B_{\nu n}^{1/\nu}$ never decreases as ν increases from 2 to k, so that we have for $\nu = 2, 3, \ldots, k$

$$(85) \qquad\qquad 1 \leqq \rho_{\nu n} \leqq \rho_{kn}^{\nu/k}.$$

F

It follows from (42) that $n^{-(\nu-2)/2}\lambda_{\nu n}$ is the νth order semi-invariant of $\mathfrak{F}_n(x)$. Further, it follows from (27) that $|\Gamma_{\nu n}| \leqq \nu^\nu B_{\nu n}$, and hence

$$(86) \qquad |\lambda_{\nu n}| \leqq \nu^\nu \rho_{\nu n} \leqq (k^k \rho_{kn})^{\nu/k}.$$

In the particular case of equal components $B_{\nu n}$, $\Gamma_{\nu n}$, $\rho_{\nu n}$ and $\lambda_{\nu n}$ are all independent of n, and we have $B_{\nu n} = \beta_\nu$, $\Gamma_{\nu n} = \gamma_\nu$, where β_ν and γ_ν denote the νth order absolute moment and the νth order semi-invariant of the common d.f. $F(x)$.

Besides the case of equal components, we shall also sometimes consider the case when the following condition is satisfied: it is possible to find two positive constants g and G such that for all n

$$(87) \qquad B_{2n} > g, \quad B_{kn} < G.$$

Obviously this case includes the case of equal components. If (87) is satisfied, it follows from (85) and (86) that $\rho_{\nu n}$ and $\lambda_{\nu n}$ are uniformly bounded for all $n \geqq 1$ and for $\nu = 2, 3, \ldots, k$.

We now consider the c.f. $\mathfrak{f}_n(t)$ of the variable $(X_1 + \ldots + X_n)/s_n$, as defined by (83). Putting

$$(88) \qquad T_{kn} = \frac{\sqrt{n}}{4\rho_{kn}^{3/k}},$$

our first object will be to show that in the interval $|t| \leqq \sqrt[3]{T_{kn}}$ there exists a certain expansion of $\mathfrak{f}_n(t)$ which, in the particular case of equal components, becomes an ordinary asymptotic expansion in powers of $n^{-\frac{1}{2}}$. (In the case of equal components, ρ_{kn} is independent of n and thus T_{kn} is, for large values of n, of the same order of magnitude as $\sqrt{n}$.)

Lemma 2.[1] *For* $|t| \leqq \sqrt[3]{T_{kn}}$ *we have*

$$(89) \qquad e^{\frac{t^2}{2}} \mathfrak{f}_n(t) = 1 + \sum_{\nu=1}^{k-3} \frac{P_{\nu n}(it)}{n^{\nu/2}} + \frac{\Theta_k}{T_{kn}^{k-2}} (|t|^k + |t|^{3(k-2)}),$$

where

$$(90) \qquad P_{\nu n}(it) = \sum_{j=1}^{\nu} c_{j\nu n}(it)^{\nu+2j}$$

is a polynomial of degree 3ν in (it), the coefficient $c_{j\nu n}$ being a poly-

---

[1] Cramér [2].

nomial in $\lambda_{3n}, \lambda_{4n}, ..., \lambda_{\nu-j+3,n}$ *with numerical coefficients, such that*

$$(91) \qquad c_{j\nu n} = \Theta_k \rho_{kn}^{\frac{\nu+2j}{k}}.$$

Thus in the case of equal components $P_{\nu n}(it)$ *is independent of* n, *while in the more general case when* (87) *is satisfied the coefficients of* $P_{\nu n}(it)$ *are bounded for all* n.

For every $r = 1, 2, ..., n$ we have by (66)

$$(92) \qquad U = f_r(t/s_n) - 1 = \sum_{\nu=2}^{k-1} \frac{\alpha_{\nu r}}{\nu!}\left(\frac{it}{s_n}\right)^\nu + \vartheta \frac{\beta_{kr}}{k!}\left(\frac{t}{s_n}\right)^k.$$

For $|t| \leq \sqrt[3]{T_{kn}}$ we obtain, however, by (84) and (88)

$$(93) \qquad \frac{\beta_{kr}^{1/k}|t|}{s_n} \leq \frac{(nB_{kn})^{1/k}}{(nB_{2n})^{1/2}}\frac{\sqrt[6]{n}}{\rho_{kn}^{1/k}} = n^{\frac{1}{k}-\frac{1}{3}} \leq 1,$$

and thus we obtain from (20)

$$|U| \leq \sum_{\nu=2}^{\infty} \frac{1}{\nu!}\left(\frac{\beta_{kr}^{1/k}|t|}{s_n}\right)^\nu \leq e-2 < \tfrac{3}{4}.$$

For $|U| < \tfrac{3}{4}$ we have, however,

$$(94) \qquad \log(1+U) = \sum_{1 \leq j < k/2} (-1)^{j+1}\frac{U^j}{j} + \Theta_k U^{k/2}.$$

According to (92) U is, formally, a polynomial in t (in reality, the factor ϑ depends of course on t), and the series

$$\sum_{\nu=2}^{\infty} \frac{1}{\nu!}\left(\frac{\beta_{kr}^{1/k}|t|}{s_n}\right)^\nu$$

is a majorating expression for this polynomial. For any power U^j, where $1 \leq j < k/2$, we thus obtain from (92) the expansion

$$U^j = \sum_{\nu=2j}^{k-1} \delta_{\nu j r}\left(\frac{it}{s_n}\right)^\nu + \vartheta \sum_{\nu=k}^{\infty} \frac{1}{\nu!}\left(\frac{j\beta_{kr}^{1/k}|t|}{s_n}\right)^\nu$$

$$= \sum_{\nu=2j}^{k-1} \delta_{\nu j r}\left(\frac{it}{s_n}\right)^\nu + \Theta_k \frac{\beta_{kr}t^k}{s_n^k},$$

with coefficients $\delta_{\nu j r}$ which are independent of t. From (92) and (94), we thus obtain an expansion of $\log f_r(t/s_n)$ in powers of it, up to the term containing $(it)^{k-1}$, and with an error term of the order t^k. According to (26), the coefficient of $(it/s_n)^\nu$ in this expansion is, however, equal to $\gamma_{\nu r}/\nu!$, so that we have for $|t| \le \sqrt[3]{T_{kn}}$

$$\log f_r(t/s_n) = \sum_{\nu=2}^{k-1} \frac{\gamma_{\nu r}}{\nu!} \left(\frac{it}{s_n}\right)^\nu + \Theta_k \frac{\beta_{kr} t^k}{s_n^k}.$$

Summing here over $r = 1, 2, ..., n$, we obtain according to (83) and (84)

$$\log \mathfrak{f}_n(t) = \sum_{\nu=2}^{k-1} \frac{n\Gamma_{\nu n}}{\nu!} \left(\frac{it}{s_n}\right)^\nu + \Theta_k \frac{nB_{kn} t^k}{s_n^k}$$

$$= -\frac{t^2}{2} + n \sum_{\nu=3}^{k-1} \frac{\lambda_{\nu n}}{\nu!} \left(\frac{it}{\sqrt{n}}\right)^\nu + \Theta_k n\rho_{kn} \left(\frac{t}{\sqrt{n}}\right)^k.$$

Substituting tz for t and dividing by z^2, we have

$$V = \log\{e^{\frac{t^2}{2}} (\mathfrak{f}_n(tz))^{\frac{1}{z^2}}\} = \sum_{\nu=1}^{k-3} \frac{\lambda_{\nu+2,n} (it)^{\nu+2}}{(\nu+2)!} \left(\frac{z}{\sqrt{n}}\right)^\nu + \Theta_k \frac{\rho_{kn} t^k}{k!} \left(\frac{z}{\sqrt{n}}\right)^{k-2}.$$

If we regard here t and n as fixed, and z as a real variable such that $|z| \le 1$, we thus have for the function $V = V(z)$ an expansion in powers of z, with an error term of the order z^{k-2}. Then obviously there is a similar expansion for the function e^V, so that we may write for $|z| \le 1$

$$(95) \qquad e^V = e^{\frac{t^2}{2}} (\mathfrak{f}_n(tz))^{\frac{1}{z^2}} = 1 + \sum_{\nu=1}^{k-3} P_{\nu n}(it) \left(\frac{z}{\sqrt{n}}\right)^\nu + R(z),$$

where $R(z) = O(z^{k-2})$ as $z \to 0$. It is then readily seen that the coefficient $P_{\nu n}(it)$ is a polynomial of degree 3ν in it, which may be put in the form (90).

According to (86), a majorating series for $V = V(z)$ is

$$(96) \qquad \Theta_k (\rho_{kn}^{1/k} |t|)^3 \frac{|z|}{\sqrt{n}} \sum_{\nu=0}^{\infty} \frac{1}{\nu!} \left(\frac{\rho_{kn}^{1/k} |tz|}{\sqrt{n}}\right)^\nu,$$

and thus V^j is, for $j = 1, 2, ..., k-2$, majorated by

$$(97) \qquad \Theta_k (\rho_{kn}^{1/k} |t|)^{3j} \left(\frac{|z|}{\sqrt{n}}\right)^j \sum_{\nu=0}^{\infty} \frac{1}{\nu!} \left(\frac{j\rho_{kn}^{1/k} |tz|}{\sqrt{n}}\right)^\nu.$$

From the development

$$e^V = \sum_{j=0}^{k-3} \frac{V^j}{j!} + \vartheta V^{k-2} e^{|V|}$$

we thus obtain, since the majorating series (96) shows that $|V| < \Theta_k$ for $|t| \le \sqrt[3]{T_{kn}}$,

$$
\begin{aligned}
R(z) &= \Theta_k \sum_{j=1}^{k-2} (\rho_{kn}^{1/k}|t|)^{3j} \left(\frac{|z|}{\sqrt{n}}\right)^j \sum_{\nu=k-2-j}^{\infty} \frac{1}{\nu!} \left(\frac{j\rho_{kn}^{1/k}|tz|}{\sqrt{n}}\right)^\nu \\
&= \Theta_k \left(\frac{\rho_{kn}^{1/k}|tz|}{\sqrt{n}}\right)^{k-2} \sum_{j=1}^{k-2} (\rho_{kn}^{1/k}|t|)^{2j} \sum_{\nu=0}^{\infty} \frac{1}{\nu!} \left(\frac{k}{\sqrt[3]{n}}\right)^\nu \\
&= \Theta_k \left(\frac{z}{\sqrt{n}}\right)^{k-2} \{(\rho_{kn}^{1/k}|t|)^k + (\rho_{kn}^{1/k}|t|)^{3(k-2)}\} \\
&= \frac{\Theta_k z^{k-2}}{T_{kn}^{k-2}} (|t|^k + |t|^{3(k-2)}).
\end{aligned}
$$

Putting $z = 1$ in (95), we thus obtain (89). Finally, the relation (91) for the coefficients $c_{j\nu n}$ follows immediately from the majorating series (97) if we observe that, in the expansion (95), a term containing the product $(it)^{\nu+2j} (z/\sqrt{n})^\nu$ can only arise from the development of the term $V^j/j!$. Thus Lemma 2 is proved.

We next consider the following Lemma 3, which gives an upper limit of $|\mathfrak{f}_n(t)|$, valid in the interval $|t| \le T_{kn}$. If the behaviour of the absolute moments β_{2n} and β_{kn} for large values of n is not too irregular, T_{kn} as defined by (88) tends to infinity with n, so that the interval $|t| \le \sqrt[3]{T_{kn}}$ of Lemma 2 is, for all sufficiently large n, contained in the interval $|t| \le T_{kn}$.

Lemma 3.[1] *For* $|t| \le T_{kn}$ *we have*

$$
|\mathfrak{f}_n(t)| \le e^{-\frac{t^2}{3}}.
$$

We have

$$
|f_r(t)|^2 = \int_{-\infty}^{\infty} \int_{-\infty}^{\infty} \cos t(x-y)\, dF_r(x)\, dF_r(y),
$$

but $\cos t(x-y) \le 1 - \frac{1}{2}t^2(x-y)^2 + \frac{1}{6}|t|^3|x-y|^3$

$$
\le 1 - \frac{1}{2}t^2(x^2 - 2xy + y^2) + \frac{2}{3}|t|^3(|x|^3 + |y|^3),
$$

---

[1] Liapounoff [2].

and thus for $|t| \leqq T_{kn}$ we obtain

$$|f_r(t)|^2 \leqq 1 - \beta_{2r}t^2 + \tfrac{4}{3}\beta_{3r}|t|^3 \leqq e^{-\beta_{2r}t^2 + \frac{4}{3}\beta_{3r}|t|^3},$$

$$|\mathfrak{f}_n(t)|^2 = \prod_{r=1}^{n} |f_r(t/s_n)|^2 \leqq e^{-t^2 + \frac{4}{3} \cdot \frac{\rho_{3n}|t|^3}{\sqrt{n}}},$$

$$|\mathfrak{f}_n(t)| \leqq e^{-\frac{t^2}{2}\left(1 - \frac{|t|}{3T_{kn}}\right)} \leqq e^{-\frac{t^2}{3}}.$$

Thus Lemma 3 is proved.

If, in the polynomial $P_{\nu n}(it)$, we replace each power $(it)^{\nu+2j}$ by $(-1)^{\nu+2j}\Phi^{(\nu+2j)}(x)$, we obtain a linear aggregate of the derivatives of the normal function $\Phi(x)$, that will be symbolically denoted by $P_{\nu n}(-\Phi)$. Thus by Lemma 2

$$(98) \qquad P_{\nu n}(-\Phi) = \sum_{j=1}^{\nu} (-1)^{\nu+2j} c_{j\nu n} \Phi^{(\nu+2j)}(x),$$

where $c_{j\nu n}$ is a polynomial in the quantities λ_{rn} such that

$$c_{j\nu n} = \Theta_k \rho_{kn}^{(\nu+2j)/k}.$$

Obviously we may write

$$\cdot(99) \qquad P_{\nu n}(-\Phi) = p_{3\nu-1,\,n}(x) e^{-\frac{x^2}{2}},$$

where $p_{3\nu-1,\,n}(x)$ is a polynomial of degree $3\nu-1$ in x. In the case of equal components, $P_{\nu n}(-\Phi)$ and $p_{3\nu-1,\,n}(x)$ are independent of n, and in the more general case when (87) is satisfied, the coefficients $c_{j\nu n}$ as well as the coefficients of $p_{3\nu-1,\,n}(x)$ are bounded for all n. According to (52) we have

$$(100) \qquad P_{\nu n}(it) e^{-\frac{t^2}{2}} = \int_{-\infty}^{\infty} e^{itx} dP_{\nu n}(-\Phi).$$

We now define two "error terms" $R_{kn}(x)$ and $r_{kn}(t)$ by writing the following expansions for the d.f. $\mathfrak{F}_n(x)$ and the c.f. $\mathfrak{f}_n(t)$

$$(101) \qquad \mathfrak{F}_n(x) = \Phi(x) + \sum_{\nu=1}^{k-3} \frac{P_{\nu n}(-\Phi)}{n^{\nu/2}} + R_{kn}(x)$$

$$= \Phi(x) + \sum_{\nu=1}^{k-3} \frac{p_{3\nu-1,\,n}(x)}{n^{\nu/2}} e^{-\frac{x^2}{2}} + R_{kn}(x),$$

$$(102) \qquad \mathfrak{f}_n(t) = e^{-\frac{t^2}{2}} + \sum_{\nu=1}^{k-3} \frac{P_{\nu n}(it)}{n^{\nu/2}} e^{-\frac{t^2}{2}} + r_{kn}(t).$$

From (100) we then obtain

$$r_{kn}(t) = \int_{-\infty}^{\infty} e^{itx} \, dR_{kn}(x).$$

Lemma 2 shows that we have $r_{kn}(t) = O(t^k)$ in the vicinity of $t = 0$, and by the argument used in IV, § 2, we conclude that

$$\int_{-\infty}^{\infty} x^\nu \, dR_{kn}(x) = 0$$

for $\nu = 0, 1, \ldots, k-1$. Thus in particular $R_{kn}(x)$ satisfies the conditions of Theorem 12.

We now proceed to the proof of the following lemma, which will be required for the proofs of Theorems 25 and 26.

Lemma 4.[1]   *For $0 < \omega < 1$, we have for all real x and all $h > 0$*

$$(103) \quad \omega \int_x^{x+h} (y-x)^{\omega-1} R_{kn}(y) \, dy = \Theta_k \left(\int_{T_{kn}}^{\infty} \frac{|\mathfrak{f}_n(t)|}{t^{\omega+1}} \, dt + \frac{1}{T_{kn}^{k-2}} \right).$$

For the proof of this lemma, we shall suppose that $T_{kn} > 1$, so that $\sqrt[3]{T_{kn}} < T_{kn}$. If this does not hold, only trivial modifications are necessary. (It will appear below that the conclusions which will be drawn from Lemma 4 are all trivial in the case $T_{kn} \leqq 1$, so that this case is not really interesting.) From (33) we obtain

$$\int_x^{x+h} (y-x)^{\omega-1} R_{kn}(y) \, dy = -\frac{1}{\pi} \Re \int_0^{\infty} \frac{r_{kn}(t)}{it} e^{-itx} \, dt \int_0^h u^{\omega-1} e^{-itu} \, du$$

---

[1] In the first edition of this Tract, Lemma 4 was stated in a different form which implies, in particular, that if the first member of (103) is replaced by

$$\omega \int_{-\infty}^x (x-y)^{\omega-1} R_{kn}(y) \, dy,$$

we obtain a relation valid for $0 < \omega \leqq k-1$. In that form, the Lemma was given by Cramér [2], and applied to the study of the asymptotic properties of certain integral averages of $\mathfrak{F}_n(x)$. (Cf. below, p. 83) and further, using the inequality (36),

$$\left| \omega \int_x^{x-h} (y-x)^{\omega-1} R_{kn}(y)\, dy \right| < C \int_0^\infty \frac{|\, r_{kn}(t)\,|}{t^{\omega+1}}\, dt$$

$$\leq C \left(\int_{T_{kn}}^\infty \frac{|\, \mathfrak{f}_n(t)\,|}{t^{\omega+1}}\, dt + A_1 + A_2 + A_3 \right),$$

where C denotes an absolute constant. For A_1, A_2 and A_3 we have, on account of Lemmas 2 and 3,

$$A_1 = \int_0^{\sqrt[3]{T_{kn}}} \frac{|\, r_{kn}(t)\,|}{t^{\omega+1}}\, dt = \frac{\Theta_k}{T_{kn}^{k-2}} \int_0^\infty \frac{t^k + t^{3(k-2)}}{t^{\omega+1}} e^{-\frac{t^2}{2}}\, dt = \frac{\Theta_k}{T_{kn}^{k-2}},$$

$$A_2 = \int_{\sqrt[3]{T_{kn}}}^{T_{kn}} \frac{|\, \mathfrak{f}_n(t)\,|}{t^{\omega+1}}\, dt < \int_{\sqrt[3]{T_{kn}}}^\infty e^{-\frac{t^2}{3}} \frac{dt}{t^{\omega+1}} < \int_{\sqrt[3]{T_{kn}}}^\infty e^{-\frac{t^2}{3}}\, dt = \frac{\Theta_k}{T_{kn}^{k-2}},$$

$$A_3 = \int_{\sqrt[3]{T_{kn}}}^\infty \frac{|\, r_{kn}(t) - \mathfrak{f}_n(t)\,|}{t^{\omega+1}}\, dt = \int_{\sqrt[3]{T_{kn}}}^\infty \left| 1 + \sum_{\nu=1}^{k-3} \frac{P_{\nu n}(it)}{n^{\nu/2}} \right| e^{-\frac{t^2}{2}} \frac{dt}{t^{\omega+1}},$$

which completes the proof of the Lemma.

4. We now consider the case $k = 3$, thus assuming the existence of finite third order absolute moments for all our random variables. From Lemmas 2 and 3 we shall then deduce the following theorem, which gives an upper limit of the modulus of the error involved in replacing $\mathfrak{F}_n(x)$ by $\Phi(x)$.

Theorem 24.[1] *Let $X_1, X_2, \ldots$ be independent variables such that X_n has the mean value zero, the s.d. σ_n and a finite third order absolute moment β_{3n}. Writing as before $s_n^2 = \sigma_1^2 + \ldots + \sigma_n^2$, the d.f. $\mathfrak{F}_n(x)$ of the variable $(X_1 + \ldots + X_n)/s_n$ satisfies for all $n \geq 1$ the inequality*

$$(104) \qquad\qquad |\mathfrak{F}_n(x) - \Phi(x)| < C \frac{\rho_{3n}}{\sqrt{n}},$$

where C is an absolute constant, while ρ_{3n} is defined by (84).

It is possible to show[2] that we can take $C = 0 \cdot 91$ in (104). Here we shall, however, restrict ourselves to giving a proof of the theorem without discussing the numerical value of C.

---

[1] This improvement of the classical inequality of Liapounoff is due to Berry and Esseen, quoted on p. 118.
[2] Zolotareff, quoted on p. 118.

Note that in the particular case of equal components $\rho_{3n} = \beta_3/\sigma^3$ is independent of n, while in the more general case when (87) is satisfied with $k=3$, ρ_{3n} is bounded for all n. Thus in both these cases the error is of the order $n^{-\frac{1}{2}}$ as $n\to\infty$. As we shall see below, after the proof of the following theorem, this order cannot be generally improved.

In order to prove the theorem we write for a fixed n

$$S = \sup | \mathfrak{F}_n(x) - \Phi(x) |$$

for all real x. Observing that $\mathfrak{F}_n(x) - \Phi(x)\to 0$ as $x\to \pm\infty$, and that $\Phi(x)$ is continuous while $\mathfrak{F}_n(x)$ is continuous to the right, it follows that it is always possible to find a number b satisfying one of the relations

$$\mathfrak{F}_n(b) - \Phi(b) = S \quad \text{or} \quad \mathfrak{F}_n(b-0) - \Phi(b) = -S.$$

Now $\mathfrak{F}_n$ and Φ are both never decreasing, and $\Phi'(x) \leq \dfrac{1}{\sqrt{2\pi}} < 1$ for all x. Taking $c = b + \frac{1}{2}S$ in the first case, and $c = b - \frac{1}{2}S$ in the second, it will be seen that throughout the interval $-\frac{1}{4}S < x < \frac{1}{4}S$, the difference $\mathfrak{F}_n(c+x) - \Phi(c+x)$ keeps a constant sign and satisfies the inequality

$$(105) \qquad | \mathfrak{F}_n(c+x) - \Phi(c+x) | \geq \tfrac{1}{4}S.$$

Further, if in Theorem 10 we take $F(x) = \epsilon(x)$, where $\epsilon(x)$ is defined by (17), the corresponding c.f. will be $f(t) = 1$, and we obtain after a change of notation

$$(106) \quad \frac{1}{\pi T} \int_{-\infty}^{\infty} \frac{1 - \cos Tx}{x^2} e^{itx}\, dx = \begin{cases} 1 - \dfrac{|t|}{T}, & |t| < T, \\ 0, & |t| \geq T, \end{cases}$$

for any $T > 0$. Thus

$$G(x) = \frac{1}{\pi T} \int_{-\infty}^{x} \frac{1 - \cos Tv}{v^2}\, dv$$

is a d.f., with the c.f. given by the second member of (106).

We now consider the convolution integral

$$R(x) = \int_{-\infty}^{\infty} [\mathfrak{F}_n(c+x-v) - \Phi(c+x-v)] \, dG(v)$$

$$= \frac{1}{\pi T} \int_{-\infty}^{\infty} \frac{1 - \cos T(x-v)}{(x-v)^2} [\mathfrak{F}_n(c+v) - \Phi(c+v)] \, dv.$$

Clearly $R(x)$ is the difference between two d.f.'s satisfying the conditions of Theorem 12. The difference between the corresponding c.f.'s is

$$r(t) = \int_{-\infty}^{\infty} e^{itx} \, dR(x) = \begin{cases} \left(1 - \dfrac{|t|}{T}\right)(\mathfrak{f}_n(t) - e^{-\frac{t^2}{2}}) e^{-cit}, & |t| < T, \\ 0, & |t| \geq T. \end{cases}$$

Since $r(t) = 0$ outside a finite interval, and $r(t) = O(t)$ as $t \to 0$, the integral (34) is convergent. By Theorem 12 we thus have, taking $x = 0$ in (35),

$$R(0) = -\frac{1}{2\pi i} \int_{-\infty}^{\infty} \frac{r(t)}{t} \, dt,$$

and hence, introducing the expressions for $R(0)$ and $r(t)$,

$$(107) \qquad \int_{-\infty}^{\infty} \frac{1 - \cos Tx}{x^2} [\mathfrak{F}_n(c+x) - \Phi(c+x)] \, dx$$

$$= \int_{-T}^{T} (T - |t|) \frac{\mathfrak{f}_n(t) - e^{-\frac{t^2}{2}}}{-2it} e^{-cit} \, dt.$$

We now take $T = T_{3n}$ as defined by (88). From Lemmas 2 (with $k = 3$) and 3 it is then easily deduced that the second member of the last relation is of modulus smaller than an absolute constant, say A. Thus we obtain from (107)

$$\left| \int_{|x| < \frac{1}{4}S} \frac{1 - \cos Tx}{x^2} [\mathfrak{F}_n(c+x) - \Phi(c+x)] \, dx \right|$$

$$< A + S \int_{|x| > \frac{1}{4}S} \frac{1 - \cos Tx}{x^2} \, dx$$

and further by (105)

$$\tfrac{1}{4}S\int_{-\infty}^{\infty} \frac{1-\cos Tx}{x^2}\,dx < A + \tfrac{5}{4}S\int_{|x|>\frac{1}{4}S} \frac{1-\cos Tx}{x^2}\,dx, \quad \frac{\pi}{4}ST < A + 10,$$

and finally for $T = T_{3n}$

$$S < \frac{16(A+10)}{\pi} \cdot \frac{\rho_{3n}}{\sqrt{n}},$$

which proves the theorem.

5. We now proceed to the case of an arbitrary $k > 3$. In this paragraph, we shall consider the particular case of equal components. It will be shown that, in this case, it is possible to give a very simple sufficient condition for the existence of an asymptotic expansion of the difference $\mathfrak{F}_n(x) - \Phi(x)$ in powers of $n^{-\frac{1}{2}}$.

In the case of equal components, the moments etc. introduced in §§ 2–3 are independent of n, so that we may write ρ_k, P_ν and $p_{3\nu-1}$ in the place of ρ_{kn}, $P_{\nu n}$ and $p_{3\nu-1,\,n}$.

We shall say that a d.f. $F(x)$ *satisfies the condition* (C) if, for the corresponding c.f. $f(t)$, we have

(C) $$\limsup_{|t|\to\infty} |f(t)| < 1.$$

By Theorem 7, the condition (C) is certainly satisfied if, in the standard decomposition of $F(x)$ according to (13), the coefficient a_I of the absolutely continuous component is different from zero. We now proceed to prove the following theorem.

Theorem 25.[1] *Let* X_1, X_2, ... *be a sequence of independent variables all having the same d.f.* $F(x)$ *with the mean value zero, the s.d.* σ, *and a finite absolute moment* β_k *of order* $k > 3$. *By* $\mathfrak{F}_n(x) = (F(\sigma x \sqrt{n}))^{n*}$ *we denote the d.f. of the variable* $(X_1 + ... + X_n)/(\sigma\sqrt{n})$. *If* $F(x)$ *satisfies the condition* (C), *we then have the expansion*

(108) $$\mathfrak{F}_n(x) = \Phi(x) + \sum_{\nu=1}^{k-3} \frac{P_\nu(-\Phi)}{n^{\nu/2}} + R_{kn}(x)$$

$$= \Phi(x) + \sum_{\nu=1}^{k-3} \frac{p_{3\nu-1}(x)}{n^{\nu/2}} e^{-\frac{x^2}{2}} + R_{kn}(x),$$

[1] Cramér [2].

with

(109) $$|R_{kn}(x)| < \frac{M}{n^{(k-2)/2}},$$

where M depends on k and on the given function F, but is independent of n and x.

Proof. From Lemma 4 we obtain, using (88) and substituting $\sigma t \sqrt{n}$ for t in the integral,

(110) $$\omega \int_x^{x+h} (y-x)^{\omega-1} R_{kn}(y)\, dy$$
$$= \Theta_k \left(\sigma^{-\omega} n^{-\omega/2} \int_{1/4(\sigma \rho_k^{3/k})}^{\infty} \frac{|\mathfrak{f}_n(\sigma t \sqrt{n})|}{t^{\omega+1}}\, dt + \frac{\rho_k^3}{n^{(k-2)/2}} \right).$$

Given any d.f. $F(x)$ satisfying the condition (C), it follows from the Remark p. 26 that we can find $c > 0$ such that $|f(t)| < e^{-c}$ for $t > 1/(4\sigma \rho_k^{3/k})$. By (83), however, $\mathfrak{f}_n(\sigma t \sqrt{n}) = (f(t))^n$, and thus we obtain from (110)

(111) $$\left| \omega \int_x^{x+h} (y-x)^{\omega-1} R_{kn}(y)\, dy \right| < M \left(\frac{e^{-cn}}{\omega} + n^{-(k-2)/2} \right).$$

M denotes here, as during the rest of this proof, an unspecified quantity depending only on k and on the given function F, but independent of n, x, h and ω.

Now $R_{kn}(y)$ is the difference between the never decreasing function $\mathfrak{F}_n(y)$ and the function $U(y) = \Phi + \sum_{\nu=1}^{k-3} n^{-\nu/2} P_\nu(-\Phi)$. The derivative $U'(y)$ obviously satisfies the relation $|U'(y)| < M$, so that we have for every y in the interval of integration

$$R_{kn}(x) - Mh < R_{kn}(y) < R_{kn}(x+h) + Mh.$$

By means of these inequalities, we obtain from (111)

$$R_{kn}(x) < M \left(h + \frac{h^{-\omega} e^{-cn}}{\omega} + h^{-\omega} n^{-(k-2)/2} \right),$$

$$R_{kn}(x+h) > -M \left(h + \frac{h^{-\omega} e^{-cn}}{\omega} + h^{-\omega} n^{-(k-2)/2} \right).$$

Replacing in the last inequality $x + h$ by x, we thus have generally

(112) $| R_{kn}(x) | < M \left(h + \dfrac{h^{-\omega} e^{-cn}}{\omega} + h^{-\omega} n^{-(k-2)/2} \right).$

Taking here $h = n^{-(k-2)/2}$, $\omega = 1/\log n$, we obtain (109), and the theorem is proved.

It is easily shown by examples that Theorem 25 does not hold true without the condition (C). Let, e.g., $F(x)$ be the step-function connected with the simple Bernoulli distribution (v, § 5):

$$F(x) = \begin{cases} 0 \text{ for } x < -p, \\ q \;\;,, \;\; -p \leqq x < q, \\ 1 \;\;,, \;\; x \geqq q. \end{cases}$$

$F(x)$ being of type II (cf. III, § 1), the condition (C) is obviously not satisfied. Taking $k = 4$, Theorem 25 would give the expansion

$$\mathfrak{F}_n(x) = \Phi(x) + \frac{p-q}{3! \sqrt{npq}} \Phi^{(3)}(x) + O\left(\frac{1}{n}\right).$$

This can, however, not be true, as it is readily seen that $\mathfrak{F}_n(x)$ has, in the vicinity of $x = 0$, discontinuities where the saltus is of the same order of magnitude as $n^{-\frac{1}{2}}$. The same example also shows that the order $n^{-\frac{1}{2}}$ given by Theorem 24 for the error of normal approximation $| \mathfrak{F}_n(x) - \Phi(x) |$ in the case of equal components cannot be generally improved.

However, it can be shown (Cramér [2], p. 56) that, even without condition (C), an asymptotic expansion of the form given in Theorem 25 holds for an appropriately *weighted average* of the function $\mathfrak{F}_n(x)$ over any given interval.

6. We shall now prove an analogue to Theorem 25 for the case of *unequal components*. We shall then have to lay down certain conditions which, roughly speaking, may be interpreted by saying that the d.f.'s of the variables X_r will be required to satisfy the condition (C) *on the average* in a certain specified sense.

According to Theorem 4, any d.f. $F_r(x)$ may be uniquely represented in the form

(113) $F_r(x) \kappa_r G_r(x) + (1 - \kappa_r) \bar{G}_r(x), \quad (0 \leqq \kappa_r \leqq 1),$

where $G_r(x)$ is a d.f. *of type I* (absolutely continuous), while $\bar{G}_r(x)$ is a d.f. which does not contain any component of type I. We now proceed to prove the following theorem.

Theorem 26. *Let X_1, X_2, ... be independent variables such that X_r has the d.f. $F_r(x)$ with the mean value zero, the s.d. σ_r, and a finite absolute moment β_{kr} of order $k > 3$. Let $F_r(x)$ be represented according to (113) and suppose that the derivative $G'_r(x)$ is of bounded total variation v_r in $(-\infty, +\infty)$. Suppose further that we have for infinitely increasing n*

$$(114) \qquad \frac{1}{\log n} \sum_{r=1}^{n} \frac{\kappa_r}{1 + v_r^2} \to \infty,$$

$$(115) \qquad \frac{T_{kn}}{s_n} \frac{1}{\log n} \sum_{r=1}^{n} \frac{\kappa_r}{1 + v_r^2} \to \infty,$$

s_n and T_{kn} being defined as in the preceding paragraphs. For the d.f. $\mathfrak{F}_n(x)$ of the variable $(X_1 + ... + X_n)/s_n$ we then have the expansion

$$\mathfrak{F}_n(x) = \Phi(x) + \sum_{\nu=1}^{k-3} \frac{P_{\nu n}(-\Phi)}{n^{\nu/2}} + R_{kn}(x)$$

$$= \Phi(x) + \sum_{\nu=1}^{k-3} \frac{p_{3\nu-1, n}(x)}{n^{\nu/2}} e^{-\frac{x^2}{2}} + R_{kn}(x),$$

with an error term $R_{kn}(x)$ satisfying the relation

$$(116) \qquad |R_{kn}(x)| < \frac{M}{T_{kn}^{k-2}},$$

where M is independent of n and x.

Remark. An important particular case is the case when (*a*) the conditions (87) are satisfied, and (*b*) the variations v_r are uniformly bounded for all $r = 1, 2, ...$. As we have

$$T_{kn}/s_n = B_{2n}/(4B_{kn}^{3/k}),$$

the conditions (114) and (115) are in this case equivalent and reduce to the single condition

$$\frac{1}{\log n} \sum_{r=1}^{n} \kappa_r \to \infty.$$

T_{kn} is in this case of the same order of magnitude as $\sqrt{n}$, so that (116, becomes

$$| R_{kn}(x) | < \frac{M}{n^{(k-2)/2}}.$$

Proof. From (91) and (98) we obtain

$$n^{-\nu/2} P_{\nu n}(-\Phi) = \Theta_k T_{kn}^{-\nu}.$$

This shows that for $T_{kn} \leqq 1$ the assertion of the theorem is trivial, so that we may assume throughout the proof $T_{kn} > 1$. From Lemma 4 we obtain, using (83), for $0 < \omega < 1$,

$$\omega \int_x^{x+k} (y-x)^{\omega-1} R_{kn}(y)\,dy = \Theta_k \left(\frac{Z}{\omega} + \frac{1}{T_{kn}^{k-2}} \right),$$

where $\quad Z = \text{upper bound of } \prod_{r=1}^{n} |f_r(t)| \text{ for } t > T_{kn}/s_n.$

Hence we obtain by the same argument as that used for the deduction of (112)

$$| R_{kn}(x) | < \Theta_k \left(h + \frac{h^{-\omega} Z}{\omega} + h^- T_{kn}^{-(k-2)} \right).$$

(For this deduction we require the result that the derivative of the function $U(t) = \Phi + \sum_{\nu=1}^{k-3} n^{-\nu/2} P_{\nu n}(-\Phi)$ satisfies, for $T_{kn} > 1$, the relation $| U'(t) | < \Theta_k$. This is easily proved by means of (91) and (98).)

Taking $h = T_{kn}^{(-k-2)}$, $\omega = 1/\log T_{kn}$, we now obtain

$$| R_{kn}(x) | < \Theta_k (T_{kn}^{-(k-2)} + Z \log T_{kn}).$$

So far we have made no use of the assumptions (114) and (115). If we can now show that, owing to these assumptions, we have for every fixed $A > 0$

(117) $$Z < \frac{M}{T_{kn}^A},$$

where M is independent of n, the theorem will obviously be proved.

By hypothesis we have, denoting by $g_r(t)$ the c.f. of $G_r(x)$,

$$|f_r(t)| \leqq \kappa_r |g_r(t)| + 1 - \kappa_r,$$

and $$|g_r(t)| = \left| -\frac{1}{it} \int_{-\infty}^{\infty} e^{itx} dG_r'(x) \right| < \frac{v_r}{|t|}.$$

For $|t| \geqq 2v_r$ we thus have

$$|f_r(t)| \leqq 1 - \tfrac{1}{2}\kappa_r,$$

and hence for $|t| < 2v_r$ by Lemma 1

$$|f_r(t)| \leqq 1 - (\kappa_r - \tfrac{1}{4}\kappa_r^2)\frac{t^2}{32v_r^2} \leqq 1 - \kappa_r \frac{t^2}{64v_r^2}.$$

It follows that we have for all $t > 0$

$$|f_r(t)| \leqq 1 - \tfrac{1}{64}\kappa_r \operatorname{Min}\left(1, \frac{t^2}{v_r^2}\right),$$

and consequently for $t > T_{kn}/s_n$

$$|f_r(t)| \leqq 1 - \tfrac{1}{64}\kappa_r \operatorname{Min}\left(1, \frac{T_{kn}^2}{s_n^2 v_r^2}\right) \leqq 1 - \frac{1}{64}\frac{\kappa_r}{1+v_r^2}\operatorname{Min}\left(1, \frac{T_{kn}^2}{s_n^2}\right)$$

$$\leqq e^{-\frac{1}{64}\frac{\kappa_r}{1+v_r^2}\operatorname{Min}\left(1, \frac{T_{kn}^2}{s_n^2}\right)},$$

$$\prod_{r=1}^{n} |f_r(t)| \leqq e^{-\frac{1}{64}\operatorname{Min}\left(1, \frac{T_{kn}^2}{s_n^2}\right)\sum_{r=1}^{n}\frac{\kappa_r}{1+v_r^2}}.$$

According to (114) and (115) the last expression is, however, for any fixed $A > 0$ and for all sufficiently large n less than $n^{-\frac{1}{2}A} < MT_{kn}^{-A}$, so that (117) holds true, and the theorem is proved.

7. It has been proved in the preceding paragraphs that, subject to certain conditions, the series[1]

$$(101a) \quad \mathfrak{F}_n(x) = \Phi(x) + \frac{P_{1n}(-\Phi)}{n^{\frac{1}{2}}} + \frac{P_{2n}(-\Phi)}{n} + \frac{P_{3n}(-\Phi)}{n^{\frac{3}{2}}} + \dots$$

gives an asymptotic expansion of $\mathfrak{F}_n(x)$ for large values of n. According to (95) and (98), the $P_{\nu n}(-\Phi)$ are for $\nu = 1, 2, \dots, k-3$ defined in the following manner. We first define an ordinary

---

[1] The formal definition of this series was given by Edgeworth [1].

polynomial $P_{vn}(t)$ by the relation

$$\frac{\sum_{v=1}^{k-3} \frac{\lambda_{v+2,n} t^{v+2}}{(v+2)!} z^v}{e^{v=1}} = 1 + \sum_{v=1}^{k-3} P_{vn}(t) z^v + O(z^{k-2}).$$

Here, λ_{vn} denotes the quantity defined by (84), so that $n^{-(v-2)/2}\lambda_{vn}$ is the vth order semi-invariant of $\mathfrak{F}_n(x)$; k is an integer such that the kth order absolute moments are known to be finite for all the components of $\mathfrak{F}_n(x)$; and finally z is an auxiliary variable which varies in the vicinity of $z = 0$. To obtain $P_{vn}(-\Phi)$ we then replace in $P_{vn}(t)$ each power t^r by the function $(-1)^r \Phi^{(r)}(x)$. In this way we obtain the expressions

$$P_{1n}(-\Phi) = -\frac{\lambda_{3n}}{3!} \Phi^{(3)}(x),$$

$$P_{2n}(-\Phi) = \frac{\lambda_{4n}}{4!} \Phi^{(4)}(x) + 10\frac{\lambda_{3n}^2}{6!} \Phi^{(6)}(x),$$

$$P_{3n}(-\Phi) = -\frac{\lambda_{5n}}{5!} \Phi^{(5)}(x) - 35\frac{\lambda_{3n}\lambda_{4n}}{7!} \Phi^{(7)}(x) - 280\frac{\lambda_{3n}^3}{9!} \Phi^{(9)}(x),$$

....

for the first terms of the development (101a). It will be remembered that in the case of equal components the λ_{vn} (and thus also the P_{vn}) are independent of n.

On the other hand, a development of the type

$$(101b) \qquad \mathfrak{F}_n(x) = \Phi(x) + \frac{c_{3n}}{3!} \Phi^{(3)}(x) + \frac{c_{4n}}{4!} \Phi^{(4)}(x) + \dots$$

has been much used by writers on mathematical statistics (cf. e.g. works by Charlier, Bruns, Gram and Thiele), and it has been claimed (without correct proof) that this expansion should possess asymptotic properties similar to those discussed above for the expansion (101a). The coefficients c_{vn} are here determined by the relation

$$c_{vn} = (-1)^v \int_{-\infty}^{\infty} H_v(x) d\mathfrak{F}_n(x),$$

where $H_v(x)$ is the vth *Hermite polynomial*:

$$H_v(x) = (-1)^v e^{\frac{x^2}{2}} \frac{d^v}{dx^v} e^{-\frac{x^2}{2}}.$$

G

From these expressions we obtain, by means of the relations
between moments and semi-invariants (IV, § 2),

$$c_{3n} = -\frac{\lambda_{3n}}{n^{\frac{1}{2}}},$$

$$c_{4n} = \frac{\lambda_{4n}}{n},$$

$$c_{5n} = -\frac{\lambda_{5n}}{n^{\frac{3}{2}}},$$

$$c_{6n} = \frac{\lambda_{6n}}{n^2} + 10\frac{\lambda_{3n}^2}{n},$$

$$c_{7n} = -\frac{\lambda_{7n}}{n^{\frac{5}{2}}} - 35\frac{\lambda_{3n}\lambda_{4n}}{n^{\frac{3}{2}}},$$

....

For larger values of n, the expressions of the $P_{\nu n}$ and the $c_{\nu n}$
become increasingly complex, but it will be seen from the above
that the two expansions (101a) and (101b) may be regarded as
rearrangements of one another. It follows from our theorems
that it is only (101a) which gives, in the ordinary sense, an
asymptotic expansion of $\mathfrak{F}_n(x)$. On the other hand, the expansion (101b) may be considered as formally simpler, since the $c_{\nu n}$
are defined by the simple relation given above, which rests on the
orthogonality properties of the Hermite polynomials.[1]

---

[1] For a more detailed analysis of the relations between the two types of expansions cf. Cramér [2].

CHAPTER VIII

A CLASS OF STOCHASTIC PROCESSES

1. In the preceding Chapters, we have been concerned with distributions of sums of the type $Z_n = X_1 + \ldots + X_n$, where the X_r are independent random variables. Z_n is then a variable depending on a discontinuous parameter n, and the passage from Z_n to Z_{n+1} means that Z_n receives the additive contribution X_{n+1}, so that we have $Z_{n+1} = Z_n + X_{n+1}$, where Z_n and X_{n+1} are independent.

Consider now the formation of Z_n by successive addition of the mutually independent contributions $X_1, X_2, \ldots$, and let us assume that each addition of a new contribution takes a finite time δ. (In a concrete interpretation the X_r might e.g. be the gains of a certain player during a series of games, every game requiring the time δ, so that Z_n is the total gain realized after n games, or after the time $n\delta$.)

The sum Z_n then arises after the time $n\delta$, and the d.f. of Z_n is thus the d.f. of the sum that has been formed during the time interval $(0, n\delta)$. Suppose now that we allow δ to tend to zero and n to tend to infinity, in such a way that $n\delta$ tends to a finite limit τ. It is conceivable that the distribution of Z_n may then tend to a definite limit, which will depend on the *continuous time parameter* τ. Thus instead of the variable Z_n with a discontinuous parameter n we should have a variable Z_τ with a continuous parameter τ, and such that the increment of Z_τ during the time interval $(\tau_1, \tau_1 + \tau_2)$ is independent of Z_{τ_1}:

$$(118) \qquad Z_{\tau_1 + \tau_2} = Z_{\tau_1} + U_{\tau_1 \tau_2},$$

where Z_{τ_1} and $U_{\tau_1 \tau_2}$ are independent.

It is, in fact, possible to give an exact meaning to the limit passage which has thus been roughly indicated. We shall, however, prefer to consider *directly* a random variable which depends on a continuous parameter and which behaves in the general way described above.[1]

2. Let τ be a continuous parameter which may be thought of as representing time. Suppose that, for every $\tau \geq 0$, we have a random variable Z_τ with the d.f. $F(x, \tau)$ and the c.f.

$$f(t, \tau) = \int_{-\infty}^{\infty} e^{itx} d_x F(x, \tau).$$

Z_0 will be supposed to be identically equal to zero, so that $F(x, 0)$ coincides with the d.f. $\epsilon(x)$ defined by (17).

The set of variables Z_τ will be said to define a *random* or *stochastic process with stationary and independent increments* (briefly: a *s.i.i. process*) if, for $\tau_1 \geq 0$, $\tau_2 > 0$, the difference $U_{\tau_1 \tau_2} = Z_{\tau_1 + \tau_2} - Z_{\tau_1}$ is a random variable which is *independent of the variable Z_{τ_1} and has a d.f. which is independent of τ_1*. We can then say that the increment of the variable Z_τ during any time interval is independent of the value assumed by the variable at the beginning of the interval, and also independent of the position of the interval on the time scale (but not, of course, independent of the *length* of the interval).

If Z_τ defines a s.i.i. process, it is seen from (118) that the d.f. of $Z_{\tau_1 + \tau_2}$ is composed by the d.f.'s of Z_{τ_1} and $U_{\tau_1 \tau_2}$. The latter d.f. is, however, by hypothesis independent of τ_1, and for $\tau_1 = 0$ we have $U_{0, \tau_2} = Z_{\tau_2} - Z_0 = Z_{\tau_2}$, so that the d.f. of $U_{\tau_1 \tau_2}$ is identical with $F(x, \tau_2)$. This gives us the following relations which may serve as an analytical definition of the s.i.i. process:

(119) $$F(x, \tau_1 + \tau_2) = F(x, \tau_1) * F(x, \tau_2),$$

(120) $$f(t, \tau_1 + \tau_2) = f(t, \tau_1) f(t, \tau_2).$$

---

[1] Particular cases of variables of this character were first studied by Bachelier [1, 2] and Lundberg [1, 2]. Further contributions were given *inter alia* by Cramér [3] and Esscher [1], in connection with the mathematical theory of insurance risk. A complete and mathematically rigorous theory, which embraces also cases much more general than the s.i.i. process, was first given by Kolmogoroff [2]. The theory of the s.i.i. process was developed by Lévy [2] under more general conditions than those considered here.

For the moments of Z_τ we shall use the notation

$$\alpha_\nu(\tau) = E(Z_\tau^\nu) = \int_{-\infty}^{\infty} x^\nu \, dF(x, \tau).$$

(Throughout the Chapter it will be understood that the variable of integration is always the *first* variable occurring in the function behind the sign d, so that we may omit the index on this sign.)

Theorem 27.[1] *Let Z_τ define a s.i.i. process, such that $\alpha_1(\tau) = 0$ and $\alpha_2(\tau)$ is finite for all $\tau > 0$. We then have*

$$(121) \quad \log f(t, \tau) = -\tfrac{1}{2}\sigma_0^2 \tau t^2 + \tau \int_{-\infty}^{\infty} \frac{e^{itx} - 1 - itx}{x^2} \, d\Omega(x),$$

where $\sigma_0^2 \geqq 0$ is a constant, and $\Omega(x)$ is a bounded and never decreasing function which is continuous at $x = 0$. Conversely, given any constant $\sigma_0^2 \geqq 0$ and any bounded and never decreasing function $\Omega(x)$ continuous at $x = 0$, (121) defines the c.f. $f(t, \tau)$ of a variable Z_τ corresponding to a s.i.i. process.

Before proceeding to the proof of this theorem, we shall consider some simple particular cases. Suppose first that $\Omega(x)$ reduces to a constant, so that the last term in the second member of (121) disappears. Then it follows from (121) that

$$F(x, \tau) = \Phi(x/(\sigma_0 \sqrt{\tau})),$$

so that Z_τ is, for every $\tau > 0$, normally distributed with the mean value 0 and the s.d. $\sigma_0 \sqrt{\tau}$. This case is often called the *Brownian movement process*, a name referring to one of its important physical applications. Suppose on the other hand $\sigma_0 = 0$ and $\Omega(x) = \lambda c^2 \epsilon(x - c)$, where $\lambda > 0$ and $c \neq 0$ are constants, and $\epsilon(x)$ is defined by (17). Then (121) gives

$$\log f(t, \tau) = \lambda \tau (e^{cit} - 1 - cit),$$

[1] Kolmogoroff [3]. Cf. also de Finetti [1, 2]. If the hypothesis $\alpha_1(\tau) = 0$ is omitted, we may apply the theorem to the variable $Z_\tau - \alpha_1(\tau)$, and choose for $\alpha_1(\tau)$ any solution (continuous or not) of the functional equation

$$\alpha_1(\tau_1 + \tau_2) = \alpha_1(\tau_1) + \alpha_1(\tau_2).$$

If we assume, e.g., that $\alpha_1(\tau)$ is *bounded* in some interval, however small, we necessarily have $\alpha_1(\tau) = c\tau$, where c is a real constant. Lévy [2] studies the s.i.i. process without assuming the existence of finite moments $\alpha_1(\tau)$ and $\alpha_2(\tau)$.

so that the variable $Z_\tau + \lambda c\tau$ has the c.f.

$$e^{\lambda\tau(e^{cit}-1)}.$$

According to (47) this corresponds to a distribution of the Poisson type. The corresponding process, which has important applications, e.g. in the theory of insurance risk, is known as the *Poisson process*.

More generally, let $\Omega(x)$ be a step-function with a finite number of steps, none of which is situated at the point $x = 0$, and put $b = \int_{-\infty}^{\infty} x^{-1} d\Omega(x)$. Then it follows from (121) that the distribution of the variable $Z_\tau + b\tau$ may be regarded as composed of one normal component (arising from the term containing σ_0) and a number of independent Poisson distributions, each of which corresponds to one step of $\Omega(x)$.

In the general case, the distribution of Z_τ is always composed of the normal component $\Phi(x/(\sigma_0 \sqrt{\tau}))$ and another component corresponding to the term containing $\Omega(x)$ in (121).

We now proceed to the proof of Theorem 27. Let us first consider the s.d. $\sqrt{\alpha_2(\tau)}$. From the fundamental relations (119) and (120) it follows that we have

$$\alpha_2(\tau_1 + \tau_2) = \alpha_2(\tau_1) + \alpha_2(\tau_2).$$

The only non-negative solution of this functional equation is, however,[1]

$$(122) \qquad \alpha_2(\tau) = \sigma^2\tau,$$

where $\sigma^2 \geqq 0$ is a constant. From (122) we deduce

$$(123) \qquad f(t, \Delta\tau) = 1 - \tfrac{1}{2}\vartheta\sigma^2 t^2 \Delta\tau$$

with $|\vartheta| \leqq 1$, so that $f(t, \Delta\tau) \to 1$ as $\Delta\tau \to 0$. According to (120) it then follows that, for every fixed t, $f(t, \tau)$ is a continuous function of τ.

From (120) we obtain further $f(t, 1/n) = \{f(t, 1)\}^{1/n}$, and hence for all rational m/n we have $f(t, m/n) = \{f(t, 1)\}^{m/n}$. By continuity this result extends immediately to all $\tau > 0$, so that we have generally

[1] Cf. Hamel [1], Hausdorff [1], p. 175.

$$(124) \qquad f(t,\tau) = \{f(t,1)\}^{\tau}.$$

According to (123), the expression

$$(125) \qquad \frac{f(t,\Delta\tau)-1}{\Delta\tau} = \frac{\{f(t,1)\}^{\Delta\tau}-1}{\Delta\tau}$$

is, for every fixed t, bounded as $\Delta\tau \to 0$. It follows that $f(t,1) \neq 0$ for all real t, and thus the expression (125) converges uniformly in every finite t-interval to the limit

$$(126) \qquad \lim_{\Delta\tau \to 0} \frac{f(t,\Delta\tau)-1}{\Delta\tau} = \log f(t,1),$$

where $\log f(t,1)$ denotes that branch of the multi-valued function which vanishes for $t=0$ and is for all real t uniquely determined by continuity.

On the other hand we have

$$(127) \qquad \frac{f(t,\Delta\tau)-1}{\Delta\tau} = \frac{1}{\Delta\tau} \int_{-\infty}^{\infty} (e^{itx} - 1 - itx)\, dF(x,\Delta\tau).$$

Putting

$$(128) \qquad H(x,\Delta\tau) = \frac{1}{\Delta\tau} \int_{-\infty}^{x} \xi^2 dF(\xi,\Delta\tau),$$

$H(x,\Delta\tau)$ is a never decreasing function of x such that

$$H(-\infty,\Delta\tau) = 0, \quad H(+\infty,\Delta\tau) = \sigma^2.$$

For every fixed $\Delta\tau > 0$, $H(x,\Delta\tau)$ is continuous at $x=0$, and we have

$$\frac{1}{\Delta\tau} \int_{-\infty}^{\infty} (e^{itx} - 1 - itx)\, dF(x,\Delta\tau) = \int_{-\infty}^{\infty} \frac{e^{itx} - 1 - itx}{x^2}\, dH(x,\Delta\tau),$$

where, for $x=0$, $(e^{itx}-1-itx)/x^2$ is to be interpreted as $-t^2/2$. According to (124), (126) and (127) we thus obtain

$$(129) \qquad \log f(t,\tau) = \tau \lim_{\Delta\tau \to 0} \int_{-\infty}^{\infty} \frac{e^{itx} - 1 - itx}{x^2}\, dH(x,\Delta\tau).$$

Consider now the function $H(x,\Delta\tau)$ for a sequence of values $\Delta_1\tau, \Delta_2\tau, \ldots$ tending to zero. It is then always possible to choose a sub-sequence $\Delta_{n_1}\tau, \Delta_{n_2}\tau, \ldots$ such that the corresponding functions $H(x, \Delta_{n_\nu} \tau)$ tend to a limit $H(x)$, in all continuity points x of the latter. From (129) we then obtain

$$(130) \qquad \log f(t,\tau) = \tau \int_{-\infty}^{\infty} \frac{e^{itx} - 1 - itx}{x^2} dH(x).$$

Obviously $H(x)$ is a never decreasing function such that

$$H(-\infty) \geqq 0, \quad H(+\infty) \leqq \sigma^2.$$

We can, however, show that in both these relations the sign of equality must hold. We obtain in fact from (130) for small values of t

$$\log f(t,\tau) = -\tfrac{1}{2}\tau t^2 \{H(+\infty) - H(-\infty)\} + o(t^2),$$

but on the other hand (122) gives

$$\log f(t,\tau) = -\tfrac{1}{2}\sigma^2 \tau t^2 + o(t^2),$$

so that we must have

$$H(-\infty) = 0, \quad H(+\infty) = \sigma^2.$$

Let, now, σ_0^2 denote the saltus of $H(x)$ at the point $x = 0$ (thus $0 \leqq \sigma_0^2 \leqq \sigma^2$) and put

$$(131) \qquad \Omega(x) = H(x) - \sigma_0^2 \epsilon(x),$$

$\epsilon(x)$ being defined by (17). Then we have

$$\Omega(-\infty) = 0, \quad \Omega(+\infty) = \sigma_1^2 = \sigma^2 - \sigma_0^2.$$

Further, $\Omega(x)$ is bounded, never decreasing and continuous at $x = 0$, and (121) follows immediately from (130), so that the first part of the theorem is proved.

The latter part of the theorem is obvious in the particular case when $\Omega(x)$ is a step-function with a finite number of steps. (Cf. the remarks made above.) Further, if $\Omega(x)$ is any function satisfying the conditions of the theorem, the second member of (121) may be uniformly approximated by means of a sequence of step-functions converging to the limit $\Omega(x)$. By Theorem 11, the corresponding d.f.'s tend to a limit which is itself a d.f., and the second member of (121) is equal to the logarithm of the c.f. of this limit. Thus (121) determines uniquely a d.f. $F(x,\tau)$, and it follows immediately from the form of (121) that the fundamental relations (119) and (120) are satisfied, so that the proof of Theorem 27 is completed.

Since $\alpha^2(\tau)$ is finite, (130) may be twice differentiated with respect to t, and we obtain

$$\int_{-\infty}^{\infty} e^{itx}\,dH(x) = -\frac{1}{\tau}\frac{\partial^2}{\partial t^2}\log f(t,\tau).$$

But $H(x)/\sigma^2$ is a d.f. which is uniquely determined by its c.f. It follows that we must reach the same limit $H(x)$ for every sequence $\Delta_1\tau, \Delta_2\tau, \ldots$ tending to zero. This implies, however, that we have $\lim_{\Delta\tau\to 0} H(x, \Delta\tau) = H(x)$ in every continuity point of $H(x)$.

This leads to an interesting interpretation of Theorem 27. For $x < 0$, we have by (128) and (131) in every continuity point of $\Omega(x)$, as $\Delta\tau\to 0$,

$$\frac{F(x,\Delta\tau)}{\Delta\tau} = \int_{-\infty}^{x}\frac{dH(\xi,\Delta\tau)}{\xi^2} \to \int_{-\infty}^{x}\frac{d\Omega(\xi)}{\xi^2} = \Pi_1(x),$$

and for $x > 0$,

$$\frac{1-F(x,\Delta\tau)}{\Delta\tau} = \int_{x}^{\infty}\frac{dH(\xi,\Delta\tau)}{\xi^2} \to \int_{x}^{\infty}\frac{d\Omega(\xi)}{\xi^2} = \Pi_2(x).$$

This may be written

$$F(x,\Delta\tau) = \Pi_1(x)\,\Delta\tau + o(\Delta\tau), \quad (x<0),$$
$$1 - F(x,\Delta\tau) = \Pi_2(x)\,\Delta\tau + o(\Delta\tau), \quad (x>0),$$

The probability that, during the infinitely small time $\Delta\tau$, a variation $< x < 0$ occurs in the value of the variable Z_τ is thus asymptotically equal to $\Pi_1(x)\Delta\tau$, while the probability of a variation $> x > 0$ is asymptotically equal to $\Pi_2(x)\Delta\tau$.

Thus the function $\Omega(x)$ determines the *discontinuous part* of the variation of Z_τ, while obviously the constant σ_0 determines the *continuous part*.[1] Further we have

$$\int_{-\infty}^{0} x^2\,d\Pi_1(x) + \int_{0}^{\infty} x^2\,|\,d\Pi_2(x)\,| = \int_{-\infty}^{\infty} d\Omega(x) = \sigma_1^2,$$
$$a_2(\tau) = \sigma^2\tau = \sigma_0^2\tau + \sigma_1^2\tau,$$

[1] It should be noted that the d.f. $F(x,\tau)$ is always continuous with respect to τ, although the variable Z_τ may suffer discontinuous changes of value, if $\Omega(x)$ is not identically zero.

so that the variance $\alpha_2(\tau)$ of Z_τ is the sum of one term due to the continuous part of the variation and one term due to the discontinuous part.

3. By means of the remarks made in §1, it will be easily understood that the s.i.i. process, as defined in §2, presents a great analogy with the "case of equal components" in the problem of addition of independent variables treated in Chapters VI–VII.[1] Roughly speaking, we are here concerned not with a *sum*, but with an *integral*, the elements of which are independent random variables (cf. Lévy [2]).

It is then fairly obvious that our previous theorems bearing on the case of equal components, such as Theorems 20 and 25, should hold, *mutatis mutandis*, also for the case of a s.i.i. process. In fact, the variable $Z_\tau/(\sigma\sqrt{\tau})$ with the d.f.

$$\mathfrak{F}(x,\tau) = F(\sigma x\sqrt{\tau},\tau)$$

and the c.f. $$\mathfrak{f}(t,\tau) = f(t/(\sigma\sqrt{\tau}),\tau)$$

is directly analogous to the previously considered variable $(X_1 + \ldots + X_n)/(\sigma\sqrt{n})$ with the d.f. $\mathfrak{F}_n(x)$ and the c.f. $\mathfrak{f}_n(t)$. Instead of the discontinuous parameter n, we are here concerned with the continuous parameter τ.

The relation (121) may be written

$$\log f(t,\tau) = -\tfrac{1}{2}\sigma^2\tau t^2 + \tau\int_{-\infty}^{\infty} \frac{e^{itx} - 1 - itx - \tfrac{1}{2}(itx)^2}{x^2}\, d\Omega(x).$$

Substituting here $t/(\sigma\sqrt{\tau})$ for t, we obtain

(132)

$$\log \mathfrak{f}(t,\tau) = -\frac{t^2}{2} + \tau\int_{-\infty}^{\infty} \frac{e^{\frac{itx}{\sigma\sqrt{\tau}}} - 1 - \dfrac{itx}{\sigma\sqrt{\tau}} - \dfrac{1}{2}\left(\dfrac{itx}{\sigma\sqrt{\tau}}\right)^2}{x^2}\, d\Omega(x).$$

---

[1] If we omit the condition laid down at the beginning of § 2 that the distribution of the increase $Z_{\tau_1+\tau_2} - Z_{\tau_1}$ should be independent of τ_1, we arrive at a more general kind of random process related to the general problem of addition of independent variables in the same way as the process here considered is related to the particular case of equal components. Subject to appropriate conditions, Theorems 27–30 can be generalized to this case. (For a generalization of Theorem 27 along these lines cf. Lévy [2], who considers also the case when $\alpha_2(\tau)$ is not finite.)

In a way which is closely similar to the proof of Theorem 20, it is now easily shown that the last term of this expression tends to zero as $\tau \to \infty$, uniformly in every finite t-interval. We thus have the following theorem directly analogous to Theorem 20.

Theorem 28.[1] *As $\tau \to \infty$, the d.f. $\mathfrak{F}(x,\tau)$ of the variable $Z_\tau/(\sigma\sqrt{\tau})$ tends to the normal function $\Phi(x)$.*

In order to obtain an asymptotic expansion of $\mathfrak{F}(x,\tau)$ for large values of τ, analogous to the expansion given by Theorem 25, we shall suppose henceforth that there is an integer $k \geq 3$, such that the absolute moment of order $k-2$ of the function $\Omega(x)$ occurring in (121) is finite. We put for $\nu = 3, \ldots, k$

(133)
$$\lambda_\nu = \frac{1}{\sigma^\nu} \int_{-\infty}^\infty x^{\nu-2}\, d\Omega(x),$$
$$\rho_\nu = \frac{1}{\sigma^\nu} \int_{-\infty}^\infty |x|^{\nu-2}\, d\Omega(x),$$
$$T_{k\tau} = \frac{\sqrt{\tau}}{4\rho_k^{3/k}}.$$

These notations are analogous to those introduced in VII, §3, by (84) and (88). We can now prove the following lemmas, which are directly analogous to Lemmas 2 and 3.

Lemma 5. *For $|t| \leq \sqrt[3]{T_{k\tau}}$ we have*
$$e^{\frac{t^2}{2}} \mathfrak{f}(t,\tau) = 1 + \sum_{\nu=1}^{k-3} \frac{P_\nu(it)}{\tau^{\nu/2}} + \frac{\Theta_k}{T_{k\tau}^{k-2}}(|t|^k + |t|^{3(k-2)}),$$

where $P_\nu(it)$ is the polynomial of degree 3ν in (it), which is obtained by replacing in the polynomial $P_{\nu n}(it)$ of Lemma 2 the quantities $\lambda_{\nu n}$ defined by (84) by the quantities λ_ν defined by (133).

Lemma 6. *For $|t| \leq T_{k\tau}$ we have*
$$|\mathfrak{f}(t,\tau)| \leq e^{-\frac{t^2}{3}}.$$

The proofs of these lemmas, which are based on the relation (132), are so closely similar to the proofs of Lemmas 2 and 3 that

[1] Lévy [2].

they need not be explicitly given here. Finally, putting in analogy to (101)

$$(134) \quad \mathfrak{F}(x,\tau) = \Phi(x) + \sum_{\nu=1}^{k-3} \frac{P_\nu(-\Phi)}{\tau^{\nu/2}} + R_k(x,\tau)$$

$$= \Phi(x) + \sum_{\nu=1}^{k-3} \frac{p_{3\nu-1}(x)}{\tau^{\nu/2}} e^{-\frac{x^2}{2}} + R_k(x,\tau),$$

where $p_{3\nu-1}(x)$ is a polynomial of degree $3\nu-1$ in x, independent of τ, we obtain in the same way as in VII, § 3 the following fundamental Lemma corresponding to Lemma 4.

Lemma 7. *For $0 < \omega < 1$, we have for all real x and all $h > 0$*

$$\omega \int_x^{x+h} (y-x)^{\omega-1} R_k(y,\tau) \, dy = \Theta_k \left(\int_{T_{k\tau}}^\infty \frac{|\mathfrak{f}(t,\tau)|}{t^{\omega+1}} \, dt + \frac{1}{T_{k\tau}^{k-2}} \right).$$

Proceeding in the same way as in VII, §§ 4–5, we can now use Lemma 7 to obtain information as to the behaviour of $\mathfrak{F}(x,\tau)$ for large values of τ. In the first place, we have the following theorem, the proof of which is directly analogous to that of Theorem 24 and need not be given here.

Theorem 29. *If the quantity ρ_3 defined by (133) is finite, we have*

$$|\mathfrak{F}(x,\tau) - \Phi(x)| < C \frac{\rho_3}{\sqrt{\tau}},$$

where C is an absolute constant.

Further, we can now prove the following theorem which gives an asymptotic expansion of $\mathfrak{F}(x,\tau)$ analogous to that obtained in Theorem 25.

Theorem 30.[1] *Suppose that the variable Z_r considered in Theorem 27 satisfies the following conditions:*

(I) *The absolute moment ρ_k as defined by (133) is finite for some integer $k > 3$;*

(II) *For some $\tau > 0$, the d.f. $F(x,\tau)$ satisfies the condition (C) of VII, § 5.*

[1] Cramér [4].

For the d.f. $\mathfrak{F}(x,\tau)$ *of the variable* $Z_\tau/(\sigma\sqrt{\tau})$, *we then have the expansion* (134) *with*

(135) $$|R_k(x,\tau)| < \frac{M}{\tau^{(k-2)/2}},$$

M being independent of τ *and* x.

Further, any of the following conditions (IIa) *and* (IIb) *is sufficient for the validity of* (II):

(IIa) $\sigma_0^2 > 0$;

(IIb) $\Omega(x) = \Omega_1(x) + \Omega_2(x)$, *where* $\Omega_1(x)$ *and* $\Omega_2(x)$ *are both never decreasing, while* $\Omega_1(x)$ *is absolutely continuous and does not reduce to a constant.*

If (II) is satisfied for a single $\tau > 0$, it follows from (121) that the same thing holds for every $\tau > 0$, and thus in particular for $\tau = 1$. From Lemma 7 we obtain according to (124)

199643

$$\omega \int_x^{x+h} (y-x)^{\omega-1} R_k(y,\tau)\,dy$$
$$= \Theta_k \left(\sigma^{-\omega}\tau^{-\omega/2} \int_{1/(4\sigma\rho_k^{3/k})}^\infty \frac{|\mathfrak{f}(t,1)|^\tau}{t^{\omega+1}}\,dt + \frac{\rho_k^3}{\tau^{(k-2)/2}} \right),$$

which corresponds to (110). By means of the condition (C) we then obtain

$$\left| \omega \int_x^{x+h} (y-x)^{\omega-1} R_k(y,\tau)\,dy \right| < M\left(\frac{e^{-c\tau}}{\omega} + \tau^{-(k-2)/2} \right),$$

and the rest of the proof of (135) is perfectly similar to the proof of Theorem 25. The last part of the theorem is easily proved by considering the real part of $\log f(t,\tau)$ according to (121).

THIRD PART

DISTRIBUTIONS IN R_k

The object of this Part is to show that many of the results obtained above for distributions in a one-dimensional space can be generalized to any number of dimensions. We shall, in the main, restrict ourselves to a brief discussion of some typical generalizations of this kind.

CHAPTER IX[1]

GENERAL PROPERTIES.
CHARACTERISTIC FUNCTIONS

1. For a distribution in a one-dimensional space, the only possible discontinuities arise from discrete points which, in terms of the mechanical interpretation used in Chapter II, are bearers of positive quantities of mass. As soon as the number of dimensions exceeds unity, the question of the discontinuities becomes, however, more complicated. Thus in a k-dimensional space, the whole mass may be concentrated to a sub-space of less than k dimensions (line, surface, ...), though there is no single point that carries a positive quantity of mass.

Given a random variable $X = (\xi_1, \ldots, \xi_k)$ in the k-dimensional space R_k, we denote as in Chapter II the corresponding pr.f. by $P(S)$ and the d.f. by $F(x_1, \ldots, x_k)$. Just as in the case $k = 1$, there can at most be a finite number of points A such that $P(A) > a > 0$, and hence at most an enumerable set of points B such that $P(B) > 0$. We shall call this set the *point spectrum* of the distribution.

---

[1] The general theory of completely additive set functions in a k-dimensional space has been developed by Radon [1], Bochner [2] and Haviland [1, 2, 3]. A comprehensive account of the principal results of the theory is given by Jessen-Wintner [1].

According to II, § 3, every component ξ_i of X is itself a random variable, and the corresponding (one-dimensional) distribution is found by projecting the original distribution on the axis of ξ_i. Let Q_i be the set of real numbers which are discontinuities of the distribution of ξ_i, and form the (at most enumerable) set $Q = Q_1 + \ldots + Q_k$. Further, let J denote a k-dimensional interval

$$a_i < \xi_i \leqq b_i,$$

and consider the probability $P(J)$ of the "event" $X \subset J$ as a function of the variables a_i and b_i. It is then obvious that, as long as no a_i and no b_i belong to the set Q, $P(J)$ is a *continuous* function of these variables.

Any interval J such that no a_i and no b_i belongs to Q will be called a *continuity interval* of the distribution. If two distributions coincide for every interval which is a continuity interval for both distributions, it follows from Theorem 2 that the corresponding d.f.'s are always equal, and thus by the same theorem the distributions are identical.

If a sequence of pr.f.'s $\{P_n(S)\}$ converges to a completely additive set function $P^*(S)$ in every continuity interval of the latter, we shall say simply that $\{P_n(S)\}$ *converges to* $P^*(S)$. The symbol $P_n(S) \to P^*(S)$ will be used only in this sense. From every sequence $\{P_n(S)\}$ it is possible[1] to choose a sub-sequence which converges in this way to a limit $P^*(S)$. Obviously we cannot in general assert that $P^*(S)$ is a probability function, as we only know that $0 \leqq P^*(R_k) \leqq 1$.

Any pr.f. can always[2] (cf. Theorem 4) be uniquely represented as a sum of three components

$$(136) \qquad P(S) = a_\mathrm{I} P_\mathrm{I}(S) + a_\mathrm{II} P_\mathrm{II}(S) + a_\mathrm{III} P_\mathrm{III}(S),$$

where a_I, a_II, a_III are non-negative numbers with the sum 1, while P_I, P_II, P_III are pr.f.'s such that

$P_\mathrm{I}(S)$ is absolutely continuous; $P_\mathrm{I}(S) = \int_S D(X) \, dX$, where

---

[1] Radon [1]. This is proved in practically the same way as in the one-dimensional case.

[2] Radon [1].

$D(X)$ is a non-negative point function in R_k which is called the *probability density* or *density function* of the distribution defined by $P_{\mathrm{I}}(S)$.

$P_{\mathrm{II}}(S)$ is purely discontinuous; $P_{\mathrm{II}}(S) = 1$ if S coincides with the point spectrum of $P(S)$.

$P_{\mathrm{III}}(S)$ is "singular"; the point spectrum of $P_{\mathrm{III}}(S)$ is empty and there exists a Borel set S of measure zero such that $P_{\mathrm{III}}(S) = 1$.

2. A real-valued function $g(X)$ which is finite and uniquely defined for all points[1] of R_k is, according to II, § 3, a random variable with a uniquely defined one-dimensional distribution. By $(15a)$ we have for the mean value of this variable the expression

$$E(g(X)) = \int_{R_k} g(X)\, dP,$$

subject to the condition that the integral is absolutely convergent. The mean values of the particular functions

$$g(X) = \xi_1^{\nu_1} \dots \xi_k^{\nu_k}, \quad (\nu_i = 0, 1, 2, \dots)$$

are called the *moments* of the distribution. We shall use the notations

$$m_i = E(\xi_i),$$

$$\mu_{ij} = E((\xi_i - m_i)(\xi_j - m_j)),$$

$$\sigma_i^2 = D^2(\xi_i) = \mu_{ii} = E((\xi_i - m_i)^2).$$

Putting

$$r_{ij} = \frac{\mu_{ij}}{\sigma_i \sigma_j},$$

it is then easily shown that we have $-1 \leqq r_{ij} \leqq 1$, and that the extreme values $r_{ij} = \pm 1$ can only be reached if, in the two-dimensional distribution of the "combined" variable (ξ_i, ξ_j), the whole mass is situated on one of the straight lines

$$(\xi_i - m_i)/\sigma_i = \pm (\xi_j - m_j)/\sigma_j.$$

r_{ij} is called the *coefficient of correlation* between ξ_i and ξ_j, and plays an important part in the statistical applications.

More generally, the quadratic form

$$\sum_{i,j} \mu_{ij} u_i u_j = \int_{R_k} \left(\sum_i u_i \xi_i \right)^2 dP$$

[1] Except possibly for certain points forming a set Σ such that $P(\Sigma) = 0$.

is *never negative*, which implies that the determinant $||\,\mu_{ij}\,||$, as well as all its principal minors, is $\geqq 0$.

3. The *characteristic function* of a distribution in R_k is the mean value

$$(137)\quad f(t_1, \ldots, t_k) = E\left(e^{i(t_1\xi_1+\ldots+t_k\xi_k)}\right) = \int_{R_k} e^{i(t_1\xi_1+\ldots+t_k\xi_k)}\,dP.$$

Unless explicitly stated otherwise, the t_i will be considered as *real* variables, so that f may be considered as a function of the real point $(t_1, \ldots, t_k)$ in R_k.

Obviously f is a uniformly continuous function of the t_i in the whole space, and we have always $|f| \leqq 1$. The generalization of Theorems 6–8 to any number of dimensions is comparatively easy, and will not be dealt with here.

If all moments up to a certain order are finite, we have for small values of $|t_i|$ an expansion of f analogous to (25). If, in particular, all μ_{ij} are finite and all m_i are equal to zero, we have

$$(138)\qquad f(t_1, \ldots, t_k) = 1 - \tfrac{1}{2}\sum_{i,j}\mu_{ij}t_it_j + o\left(\sum_i t_i^2\right).$$

We shall now consider the generalization of Theorem 9, and for the sake of formal simplicity we take first the case of a two-dimensional space R_2. The generalized theorem will then be as follows: If the interval J defined by

$$x_1 < \xi_1 \leqq x_1+h_1,\quad x_2 < \xi_2 \leqq x_2+h_2$$

is a continuity interval of the distribution, we have

$$(139)\quad P(J) = F(x_1+h_1, x_2+h_2) - F(x_1+h_1, x_2)$$
$$- F(x_1, x_2+h_2) + F(x_1, x_2)$$
$$= \lim_{T\to\infty}\frac{1}{4\pi^2}\int_{-T}^{T}\int_{-T}^{T}\frac{1-e^{-it_1h_1}}{it_1}\frac{1-e^{-it_2h_2}}{it_2}e^{-i(t_1x_1+t_2x_2)}f(t_1,t_2)\,dt_1dt_2.$$

We have, in fact, for the quantity behind the sign $\lim_{T\to\infty}$, the expression

$$\int_{R_2}\psi_1(\xi_1)\psi_2(\xi_2)\,dP,$$

H

where

$$\psi_i(\xi_i) = \frac{1}{\pi}\int_0^T \frac{\sin t\,(\xi_i - x_i)}{t}\,dt - \frac{1}{\pi}\int_0^T \frac{\sin t\,(\xi_i - x_i - h_i)}{t}\,dt.$$

As $T \to \infty$, the product $\psi_1(\xi_1)\psi_2(\xi_2)$ tends to unity for every interior point (ξ_1, ξ_2) of J, and to zero for every point (ξ_1, ξ_2) outside J.

It is then obvious that the proof of (139) can be performed by an easy extension of the argument used in the proof of Theorem 9. As in the case of Theorem 9, we deduce immediately the corollary that a two-dimensional d.f. is uniquely determined by its c.f.

It is clear that the argument is perfectly general, so that we may state the following theorem.

Theorem 9 a.[1] *If the k-dimensional interval J defined by $x_i < \xi_i \leqq x_i + h_i$ $(i = 1, 2, \ldots, k)$ is a continuity interval for the probability function $P(S)$, we have*

$$P(J) = \lim_{T \to \infty} \frac{1}{(2\pi)^k} \int_{-T}^T \cdots \int_{-T}^T \frac{1 - e^{-it_1 h_1}}{it_1} \cdots \frac{1 - e^{-it_k h_k}}{it_k}$$
$$\times e^{-i(t_1 x_1 + \ldots + t_k x_k)} f(t_1, \ldots, t_k)\,dt_1 \ldots dt_k.$$

Hence it follows that a probability distribution in R_k is uniquely determined by its characteristic function.[2]

4. Before proceeding to further generalizations of a similar kind, we shall now introduce a method of induction[3] which can often be used for the extension of theorems on one-dimensional distributions to any number of dimensions.

Consider a random variable $X = (\xi_1, \ldots, \xi_k)$ in R_k with the pr.f. $P(S)$. Let $T = (t_1, \ldots, t_k)$ denote a fixed point in R_k such that $T \neq (0, \ldots, 0)$, and consider the one-dimensional random variable

$$U = t_1 \xi_1 + \ldots + t_k \xi_k.$$

The c.f. of this variable is by (137)

(140) $E(e^{itU}) = E(e^{it(t_1 \xi_1 + \ldots + t_k \xi_k)}) = f(tt_1, \ldots, tt_k).$

This is a relation between the c.f. of a k-dimensional variable X

---

[1] Romanovsky [1], Haviland [3].
[2] There is also a similar straightforward generalization of Theorem 10 which will not be explicitly given here.
[3] Cramér-Wold [1].

and the c.f. of a certain associated one-dimensional variable U. Since both t and $t_1, \ldots, t_k$ are arbitrary, it will in many cases be possible to use this relation for the purpose in view.

Denoting by $S_{T,x}$ the *half-space* defined by the inequality

$$(141) \qquad U = t_1 \xi_1 + \ldots + t_k \xi_k \leqq x,$$

we observe that $P(S_{T,x})$, considered as a function of the real variable x, is the d.f. of the random variable U. From (140), we now obtain in the first place the following theorem, the one-dimensional case of which is, of course, trivial.

Theorem 31.[1] *If two probability functions in R_k coincide for every half-space $S_{T,x}$, they are identical.*

In order to prove this theorem it is sufficient to remark that, by hypothesis, the associated variable U has one and the same pr.f. in both cases. Thus in the relation (140) the first member, being the c.f. of U, has the same value in both cases. Putting $t = 1$, it then follows that the c.f.'s of both distributions coincide for $T \neq (0, \ldots, 0)$. For $T = (0, \ldots, 0)$, both c.f.'s assume the value 1. Thus the c.f.'s are always equal, and then by Theorem 9a the corresponding distributions are identical.

5. We now proceed to the generalization of the important Theorem 11.

Theorem 11a.[2] *Let $\{P_n(S)\}$ be a sequence of pr.f.'s in R_k, and $\{f_n(t_1, \ldots, t_k)\}$ the corresponding sequence of c.f.'s. A necessary and sufficient condition for the convergence of $\{P_n(S)\}$ to a pr.f. $P(S)$, in every continuity interval of the latter, is that the sequence $\{f_n(t_1, \ldots, t_k)\}$ converges for every $T = (t_1, \ldots, t_k)$ to a limit $f(t_1, \ldots, t_k)$, which is continuous at the point $T = (0, \ldots, 0)$.*

When this condition is satisfied, the limit $f(t_1, \ldots, t_k)$ is identical with the c.f. of $P(S)$, and $\{f_n\}$ converges to f uniformly in every finite interval.

That the condition is necessary is proved by a straightforward

---

[1] Cramér-Wold [1].

[2] Romanovsky [1], Bochner [2], Haviland [3], Cramér-Wold [1].

generalization of the argument used in the one-dimensional case. It thus only remains (cf. the proof of Theorem 11) to prove that, if $f_n(t_1, \ldots, t_k)$ converges to a limit $f^*(t_1, \ldots, t_k)$, uniformly in $|t_i| < a$, then $P_n(S) \to P(S)$, where $P(S)$ is a pr.f.

Let $T = (t_1, \ldots, t_k)$ be a given point in R_k such that $T \neq (0, \ldots, 0)$, and consider the sequence $f_n(tt_1, \ldots, tt_k)$, where t is a real variable. By hypothesis this converges for all t to a limit, whⅰch is continuous at $t = 0$. According to the preceding paragraph, $f_n(tt_1, \ldots, tt_k)$ is, however, the c.f. of the d.f. $P_n(S_{T,x})$. Thus by Theorem 11 we have $P_n(S_{T,x}) \to F_T(x)$ in every continuity point of $F_T(x)$, where $F_T(x)$ is a d.f.

From $\{P_n(S)\}$, we now choose a sub-sequence which converges to a limit $P^*(S)$, in every continuity interval of the latter. Then it follows from the above that, in every continuity point of $F_T(x)$, we have $P^*(S_{T,x}) = F_T(x)$. Allowing here x to tend to infinity, it follows that $P^*(R_k) = 1$, and thus $P^*(S)$ is a pr.f., which we denote by $P(S)$.

In exactly the same way as in the proof of Theorem 11 we can now show that every convergent sub-sequence of $\{P_n(S)\}$ converges to the same limit $P(S)$. This is, however, equivalent to the statement that the sequence $\{P_n(S)\}$ converges to $P(S)$. Thus Theorem 11 a is proved.

We shall not enter here upon the question of a k-dimensional generalization of Theorem 12.

6. Let us consider two mutually independent variables X_1 and X_2 in R_k. The pr.f.'s will be denoted by P_1 and P_2, and the c.f.'s by f_1 and f_2 respectively. The sum $X_1 + X_2$, formed according to the ordinary rule of vector addition, is a k-dimensional vector function of the combined variable (X_1, X_2), and thus according to II, §6 (cf. v, §1), $X_1 + X_2$ is a random variable in R_k, with a probability distribution uniquely determined by P_1 and P_2. We shall now prove the following theorem, which corresponds to Theorem 13.

Theorem 13a.[1] *If X_1 and X_2 are mutually independent random variables in R_k with the pr.f.'s P_1 and P_2, and the c.f.'s f_1 and f_2, then the sum $X_1 + X_2$ has the c.f.*

$$(142) \qquad f(t_1, ..., t_k) = f_1(t_1, ..., t_k) f_2(t_1, ..., t_k),$$

and the pr.f.

$$(14 \ ) \qquad P(S) = \int_{R_k} P_1(S - X) dP_2 = \int_{R_k} P_2(S - X) dP_1,$$

where $S - X$ denotes the set of all points $\mathfrak{X} - X$, where $\mathfrak{X}$ belongs to S.

As in the one-dimensional case, the relation (142) for the c.f.'s is an immediate consequence of the definition of the c.f. according to (137).

Let us now consider the second member of (143). For every Borel set S, $P_1(S - X)$ is a bounded, non-negative and B-measurable function of the point X, so that according to I, § 3, the integral always exists.[2] Obviously the value $P(S)$ of this integral is a completely additive non-negative function of the set S which, for $S = R_k$, assumes the value 1, i.e. a pr.f. If, now, we can show that the c.f. of $P(S)$, which we denote by $\bar{f}$, is identical with f as given by (142), it follows that $P(S)$ must be the pr.f. of the variable $X_1 + X_2$, and then by reasons of symmetry $P(S)$ will also be equal to the third member of (143), so that the theorem will be proved.

If the set S is a half-space $S_{T,x}$ as defined by (141), it follows from the expression of $P(S)$ as an integral that the one-dimensional d.f. $P(S_{T,x})$ is the convolution of $P_1(S_{T,x})$ and $P_2(\dot{S}_{T\ x})$, or in the notation of v, § 1,

$$P(S_{T,x}) = P_1(S_{T,x}) * P_2(S_{T,x}).$$

---

[1] Bochner [2], Haviland [2], Cramér-Wold [1].

[2] This may be shown in the following way. If, in particular, S is an interval of the type $S_{x_1, ..., x_k}$ considered in II, § 2, we have, putting

$$X = (\xi_1, ..., \xi_k), \quad P_1(S - X) = F_1(x_1 - \xi_1, ..., x_k - \xi_k),$$

where F_1 is the d.f. corresponding to the pr.f. P_1. Thus in this case P_1, regarded as a function of X, is bounded, non-negative and B-measurable. From the complete additivity of $P_1(S - X)$ with respect to S it then follows that the same properties must hold for every Borel set S in R_k.

Thus according to Theorem 13 the c.f. of the first member is the product of the c.f.'s of both components in the second member, which gives according to (140)

$$\bar{f}(tt_1, \ldots, tt_k) = f_1(tt_1, \ldots, tt_k) f_2(tt_1, \ldots, tt_k).$$

Putting here $t = 1$, we obtain the desired result.

As in the one-dimensional case, we shall say that $P(S)$ is the convolution of the components $P_1(S)$ and $P_2(S)$, and we shall use the abbreviation

$$(143a) \qquad\qquad P = P_1 * P_2 = P_2 * P_1.$$

For the sum $X_1 + X_2 + \ldots + X_n$ of n mutually independent random variables in R_k we have the pr.f.

$$P = P_1 * P_2 * \ldots * P_n,$$

and the c.f. $\qquad\qquad f = f_1 f_2 \ldots f_n.$

CHAPTER X

THE NORMAL DISTRIBUTION AND
THE CENTRAL LIMIT THEOREM

1. In order to generalize the *normal distribution* to the space R_k, it is convenient to begin with a discussion of the characteristic functions. For a normally distributed one-dimensional variable with the mean value m and the s.d. σ, we have, according to VI, §1, the c.f.

$$f(t) = e^{mit - \frac{1}{2}\sigma^2 t^2}.$$

A perfectly natural generalization of this expression to k variables is obtained by putting

$$(144) \qquad f(t_1, \ldots, t_k) = e^{i\sum_r m_r t_r - \frac{1}{2}\sum_{r,s} \mu_{rs} t_r t_s}.$$

Assuming that $f(t_1, \ldots, t_k)$ as defined by this expression is the c.f. of a probability distribution in R_k, it is seen by expansion in MacLaurin's series that m_r and μ_{rs} are the first and second order moments introduced in IX, §2. We have, in fact, the following theorem.

Theorem 32. *For any real m_r and $\mu_{rs} = \mu_{sr}$ such that the quadratic form $Q = \sum_{r,s} \mu_{rs} t_r t_s$ is never negative, $f(t_1, \ldots, t_k)$ as defined by (144) is the c.f. of a probability distribution in R_k, which will be called a normal distribution. The following two cases may occur:*

(A) If the form Q is definite positive, the corresponding distribution is absolutely continuous and will be called a proper normal distribution. The density function of this distribution is

$$(145) \qquad D(X) = D(\xi_1, \ldots, \xi_k) = \frac{1}{(2\pi)^{k/2}\sqrt{\Delta}} e^{-\frac{1}{2}q(\xi_1, \ldots, \xi_k)},$$

where $\Delta = ||\mu_{rs}|| > 0$ and $q = \sum_{r,s} \frac{\Delta_{rs}}{\Delta}(\xi_r - m_r)(\xi_s - m_s)$ is the reciprocal form of Q, with the variables $\xi_r - m_r$.

(B) *If the form Q is only semi-definite, the distribution is of the singular type and will be called an improper normal distribution. For this distribution, the whole mass is situated in a certain subspace of less than k dimensions, defined by one or more linear relations between the ξ_r (straight line, plane, hyperplane). Every improper normal distribution may be represented as the limit of a sequence of proper normal distributions.*[1]

In the case (A) we have to show that (145) is a density function, the c.f. of which is identical with (144). Consider the integral

$$G(u_1, \ldots, u_k) = \frac{1}{(2\pi)^{k/2}\sqrt{\Delta}} \int_{R_k} e^{\sum_r u_r \xi_r - \frac{1}{2}q} \, d\xi_1 \ldots d\xi_k,$$

where, until further notice, $u_1, \ldots, u_k$ are *real* variables. By means of the substitution

$$\xi_r - m_r = \sum_s \mu_{rs}(\eta_s + u_s),$$

$$\eta_r + u_r = \sum \frac{\Delta_{rs}}{\Delta}(\xi_s - m_s),$$

we obtain

$$\sum_r u_r \xi_r - \frac{1}{2}\sum_{r,s} \frac{\Delta_{rs}}{\Delta}(\xi_r - m_r)(\xi_s - m_s)$$
$$= \sum_r m_r u_r + \frac{1}{2}\sum_{r,s} \mu_{rs} u_r u_s - \frac{1}{2}\sum_{r,s} \mu_{rs}\eta_r\eta_s,$$

and hence

(146) $\quad G = e^{\sum_r m_r u_r + \frac{1}{2}\sum_{r,s} \mu_{rs} u_r u_s} \dfrac{\sqrt{\Delta}}{(2\pi)^{k/2}} \displaystyle\int_{R_k} e^{-\frac{1}{2}\sum_{r,s} \mu_{rs}\eta_r\eta_s} \, d\eta_1 \ldots d\eta_k$

$\qquad\qquad = e^{\sum_r m_r u_r + \frac{1}{2}\sum_{r,s} \mu_{rs} u_r u_s}.$

For $u_r = 0$ we obtain $G = 1$, which shows that (145) is a density function. Since both members of (146) are integral functions of the u_r, (146) holds also for complex values of the u_r. Substituting it_r for u_r, G becomes the c.f. corresponding to the

---

[1] The distinction of proper and improper normal distributions is, of course, relative to the space R_k in which the distributions are considered. A proper normal distribution in R_k becomes an improper distribution as soon as it is considered as a distribution in a space R_K with $K > k$.

density function (145), and (146) shows that this is identical with (144).

In order to prove the case (B) of Theorem 32, it is sufficient to consider a sequence of proper normal distributions, allowing the μ_{rs} to tend to the coefficients of a semi-definite form Q, while the m_r are being kept constant. Obviously the corresponding c.f.'s converge uniformly to a limit of the form (144) with a semi-definite Q, and then Theorem 11a shows that this limit is the c.f. of a certain probability distribution in R_k. The determinant Δ is, of course, equal to zero for the limiting distribution, and it then follows from (145) that the whole mass is concentrated in the set of points defined by[1] $\sum_{r,s} \Delta_{rs}(\xi_r - m_r)(\xi_s - m_s) = 0$. Now the determinant $||\Delta_{rs}^{\bullet}||$ is zero, so that this relation is equivalent to a certain number $k_1 < k$ of linear relations between $\xi_1, \ldots, \xi_k$.

2. We now proceed to the generalization of Theorems 17–20. From the expression (144) of the c.f., the following theorem is immediately deduced.

Theorem 17a. *The sum of two independent and normally distributed variables in R_k is itself normally distributed. If at least one of the components has a proper normal distribution, the same holds true for the sum.*

It should be noted that, of course, the sum may have a proper normal distribution even if both components have improper distributions, since the sum of two semi-definite forms Q may well be a definite form.

If σ is a positive quantity, and $M = (m_1, \ldots, m_k)$ is a point in R_k, we denote by $(S-M)/\sigma$ the set of all points $(X-M)/\sigma$, where X belongs to S. If $P(S)$ is the pr.f. of a random variable X in R_k, it is then clear that $P((S-M)/\sigma)$ is the pr.f. of the variable $M + \sigma X$.

Theorem 18a. *Let $P(S)$ be a pr.f. in R_k with finite second order moments. If, to any points M_1, M_2 in R_k and any positive*

---

[1] In order to avoid trivial difficulties, we assume here that there is at least one minor $\Delta_{rs} \neq 0$.

constants σ_1, σ_2, we can find M and σ such that

$$P\left(\frac{S-M_1}{\sigma_1}\right) * P\left(\frac{S-M_2}{\sigma_2}\right) = P\left(\frac{S-M}{\sigma}\right),$$

then $P(S)$ is a normal pr.f.

Theorem 19a. *If the sum of two independent variables in R_k is normally distributed, then each variable is itself normally distributed.*

Theorem 20a. *Let $P(S)$ be a pr.f. in R_k such that the first order moments m_r are all equal to zero while the second order moments μ_{rs} are finite. If X_1, X_2, ... are independent variables all having the pr.f. $P(S)$, then the pr.f. $(P(S\sqrt{n}))^{n*}$ of the variable*

$$(X_1 + \ldots + X_n)/\sqrt{n}$$

tends, as $n \to \infty$, to that normal pr.f. which has the same first and second order moments as $P(S)$.

As in the one-dimensional case, we prove first Theorem 20a, from which Theorem 18a is deduced as a corollary. We then prove Theorem 19a by means of the induction method indicated in IX, § 4.

The c.f. of $(P(S\sqrt{n}))^{n*}$ is $(f(t_1/\sqrt{n}, \ldots, t_k/\sqrt{n}))^n$. From the relation (138) it then follows, in the same way as in the one-dimensional case, that this tends to the limit

$$e^{-\frac{1}{2}\sum\limits_{r,s}\mu_{rs}t_r t_s}$$

as $n \to \infty$, uniformly in every finite interval. According to Theorem 11a, this proves Theorem 20a. Hence Theorem 18a is deduced in a perfectly similar way as in the one-dimensional case.

In order to prove Theorem 19a, we suppose that

$$P_1(S) * P_2(S) = P(S),$$

where $P(S)$ is a normal pr.f., and thus have for the corresponding c.f.'s the relation

(147) $\qquad f_1 f_2 = f = e^{iL - \frac{1}{2}Q},$

where L is a linear form and Q a non-negative quadratic form in the variables t_r.

Consider now the one-dimensional d.f.'s $P_1(S_{T,x})$, $P_2(S_{T,x})$, $P(S_{T,x})$, where $S_{T,x}$ is the half-space defined by (141). By (140) the corresponding c.f.'s are $f_1(tt_1, ..., tt_k)$, $f_2(tt_1, ..., tt_k)$ and $f(tt_1, ..., tt_k) = e^{iLt - \frac{1}{2}Qt^2}$. Thus $P(S_{T,x})$ is a normal d.f., and since according to (147) we have

$$P_1(S_{T,x}) * P_2(S_{T,x}) = P(S_{T,x}),$$

it follows from Theorem 19 that $P_1(S_{T,x})$ and $P_2(S_{T,x})$ are both normal. We thus have

$$(148) \qquad f_1(tt_1, ..., tt_k) = e^{iL_1 t - \frac{1}{2}Q_1 t^2},$$

where L_1 and Q_1 are functions of $t_1, ..., t_k$.

It follows from (140) that $f_1(tt_1, ..., tt_k)$ is the c.f. of the variable $U = t_1\xi_1 + ... + t_k\xi_k$, if we put $X_1 = (\xi_1, ..., \xi_k)$. The nth order moment of U is thus a homogeneous polynomial of order n in the t_r. Thus in (148) L_1 is a linear form and Q_1 is a quadratic form in the t_r. Since (148) is a c.f. in t, the form Q_1 must be non-negative. Putting $t = 1$ in (148), it then follows that $f_1(t_1, ..., t_k)$ is the c.f. of a normal distribution in R_k. The same holds, of course, for $f_2(t_1, ..., t_k)$, and thus Theorem 19a is proved.

3. Theorem 20a constitutes the simplest case of the *Central Limit Theorem* for random variables in R_k. It is possible to find also k-dimensional analogues of the more general theorems proved in Chapter VI, §§ 3–6, and of the theorems on asymptotic expansions, etc. given in Chapters VII–VIII. We shall content ourselves here with giving the statement of a theorem which corresponds to Theorem 21, though it does not represent a complete generalization of what has been proved in the one-dimensional case.

Theorem 21a.[1] *Let X_1, X_2, ... be a sequence of independent random variables in R_k such that every X_n has the pr.f. $P_n(S)$ with vanishing first order moments and finite second order moments*

---

[1] Theorems of a similar kind have been given by Bernstein [1], Castelnuovo [1] and Khintchine [1, 2].

$\mu_{rs}^{(n)}$. *Suppose that, as $n \to \infty$, the following two conditions are satisfied:*

$$(149) \qquad \frac{1}{n} \sum_{\nu=1}^{n} \mu_{rs}^{(\nu)} \to \mu_{rs}, \quad (r, s = 1, 2, \ldots, k),$$

where the μ_{rs} are not all equal to zero, and

$$(150) \qquad \frac{1}{n} \sum_{\nu=1}^{n} \int_{|X| > \epsilon \sqrt{n}} |X|^2 dP_{\nu} \to 0$$

for every $\epsilon > 0$, where $|X|$ denotes $\sqrt{\xi_1^2 + \ldots + \xi_k^2}$.

Then the pr.f. of the variable $(X_1 + \ldots + X_n)/\sqrt{n}$ converges to that normal pr.f. which has the first order moments zero and the second order moments μ_{rs}.

This theorem can be proved by a direct generalization of the proof of Theorem 21, which, of course, requires a little more calculation than in the one-dimensional case, but does not involve any new difficulty of principle. Obviously the condition (150) is analogous to the Lindeberg condition (64). It should be observed that the limiting distribution may well be an *improper* normal distribution, viz. if the corresponding form Q is semi-definite. Obviously this may occur even in a case when all the functions $P_n(S)$ are absolutely continuous.

BIBLIOGRAPHY

The following list contains only works actually referred to in the text.

[1] BACHELIER, L. Théorie de la spéculation. *Annales École Norm. Sup.* 17 (1900), 21.

[2] BACHELIER, L. *Calcul des probabilités.* Paris, 1912.

[1] BERNSTEIN, S. Sur l'extension du théorème limite du calcul des probabilités aux sommes de quantités dépendantes. *Math. Annalen,* 97 (1927), 1–59.

[1] BESICOVITCH, A. S. *Almost periodic functions.* Cambridge, 1932.

[1] BOCHNER, S. *Vorlesungen über Fouriersche Integrale.* Leipzig, 1932.

[2] BOCHNER, S. Monotone Funktionen, Stieltjessche Integrale und harmonische Analyse. *Math. Annalen,* 108 (1933), 378–410.

[1] CANTELLI, F. P. La tendenza ad un limite nel senzo del calcolo delle probabilità. *Rend. Circ. Mat. Palermo,* 16 (1916), 191–201.

[2] CANTELLI, F. P. Una teoria astratta del calcolo delle probabilità. *Giorn. Ist. Ital. Attuari,* 3 (1932), 257–265.

[1] CASTELNUOVO, G. *Calcolo delle probabilità.* Second ed. Bologna, 1926–28.

[1] CRAMÉR, H. Das Gesetz von Gauss und die Theorie des Risikos. *Skand. Aktuarietidskr.* 6 (1923), 209–237.

[2] CRAMÉR, H. On the composition of elementary errors. *Skand. Aktuarietidskr.* 11 (1928), 13–74 and 141–180.

[3] CRAMÉR, H. On the mathematical theory of risk. *Skandia-festskrift,* Stockholm, 1930.

[4] CRAMÉR, H. Sur les propriétés asymptotiques d'une classe de variables aléatoires. *C.R. Acad. Sci. Paris,* 201 (1935), 441–443.

[5] CRAMÉR, H. Über eine Eigenschaft der normalen Verteilungs-funktion. *Math. Zeitschrift,* 41 (1936), 405–414.

[1] CRAMÉR, H. and WOLD, H. Some theorems on distribution functions. *Journ. London Math. Soc.* 11 (1936), 290–294.

[1] EDGEWORTH, F. Y. The law of error. *Camb. Phil. Soc. Proc.* 20 (1905), 36–141.

[1] ELDERTON, W. P. *Frequency curves and correlation.* Second ed. London, 1927.

[1] ESSCHER, F. On the probability function in the collective theory of risk. *Skand. Aktuarietidskr.* 15 (1932), 175–195.

[1] FELLER, W. Über den zentralen Grenzwertsatz der Wahrscheinlich-keitsrechnung. *Math. Zeitschrift,* 40 (1935), 521–559.

[2] FELLER, W. Über den zentralen Grenzwertsatz der Wahrscheinlich-keitsrechnung, II. *Math. Zeitschrift,* 42 (1937).

[1] DE FINETTI, B. Sulle funzioni a incremento aleatorio. *Rend. R. Accad. Lincei,* (6), 10 (1929), 163–168.

[2] DE FINETTI, B. Le funzioni caratteristiche di legge istantanea. *Rend. R. Accad. Lincei*, (6), 12 (1930), 278–282.

[1] FRÉCHET, M. Sur la convergence en probabilité. *Metron*, 8 (1930), 1–48.

[2] FRÉCHET, M. *Recherches théoriques modernes. Traité du calcul des probabilités*, par E. Borel, tome I, fasc. 3, Paris, 1937.

[1] GLIVENKO, V. Sul teorema limite della teoria delle funzioni caratteristiche. *Giorn. Ist. Ital. Attuari*, 7 (1936), 160–167.

[1] HAMEL, G. Eine Basis aller Zahlen und die unstetigen Lösungen der Funktionalgleichung $f(x+y) = f(x)+f(y)$. *Math. Annalen*, 60 (1905), 459–462.

[1] HARDY, G. H., LITTLEWOOD, J. E. and PÓLYA, G. *Inequalities*. Cambridge, 1934.

[1] HAUSDORFF, F. *Mengenlehre*. Second ed. Berlin-Leipzig, 1927.

[1] HAVILAND, E. K. On distribution functions and their Laplace-Fourier transforms. *Proc. National Acad. Sci.* 20 (1934), 50–57.

[2] HAVILAND, E. K. On the theory of absolutely additive distribution functions. *Amer. Journ. Math.* 56 (1934), 625–658.

[3] HAVILAND, E. K. On the inversion formula for Fourier-Stieltjes transforms in more than one dimension. *Amer. Journ. Math.* 57 (1935), 94–100 and 382–388.

[1] HOBSON, E. W. *The theory of functions of a real variable*. Vol. I, third ed. 1927, Vol. II, second ed. 1926.

[1] JESSEN, B. and WINTNER, A. Distribution functions and the Riemann zeta function. *Trans. Amer. Math. Soc.* 38 (1935), 48–88.

[1] KEYNES, J. M. *A treatise on probability*. London, 1921.

[1] KHINTCHINE, A. Begründung der Normalkorrelation nach der Lindebergschen Methode. *Nachr. Forschungsinst. Moskau*, 1 (1928).

[2] KHINTCHINE, A. *Asymptotische Gesetze der Wahrscheinlichkeitsrechnung*. Berlin, 1933.

[3] KHINTCHINE, A. Sul dominio di attrazione della legge di Gauss. *Giorn. Ist. Ital. Attuari*, 6 (1935), 378–393.

[4] KHINTCHINE, A. Su una legge dei grandi numeri generalizzata. *Giorn. Ist. Ital. Attuari*, 7 (1936), 365–377.

[1] KOLMOGOROFF, A. Bemerkungen zu meiner Arbeit "Über die Summen zufälliger Grössen". *Math. Annalen*, 102 (1929), 484–488.

[2] KOLMOGOROFF, A. Über die analytischen Methoden in der Wahrscheinlichkeitsrechnung. *Math. Annalen*, 104 (1931), 415–458.

[3] KOLMOGOROFF, A. Sulla forma generale di un processo stocastico omogeneo. *Rend. R. Accad. Lincei*, (6), 15 (1932), 805–808 and 866–869.

[4] KOLMOGOROFF, A. *Grundbegriffe der Wahrscheinlichkeitsrechnung*. Berlin, 1933.

[1] LAGRANGE, J. L. Mémoire sur l'utilité de la méthode de prendre le milieu entre les résultats de plusieurs observations. *Misc. Taurinensia*, 5 (1770–73), 167–232. *Œuvres*, 2, Paris, 1868.

[1] LAPLACE, P. S. *Théorie analytique des probabilités*. First ed. 1812, second ed. 1814, third ed. 1820.

[1] LEBESGUE, H. *Leçons sur l'intégration.* Second ed. Paris, 1928.

[1] LÉVY, P. *Calcul des probabilités.* Paris, 1925.

[2] LÉVY, P. Sur les intégrales dont les élements sont des variables aléatoires indépendantes. *Annali R. Sci. Norm. Sup. Pisa,* (2), 3 (1934), 337–366.

[3] LÉVY, P. Propriétés asymptotiques des sommes de variables aléatoires indépendantes ou enchaînées. *Journ. Math. pures appl.* (7), 14 (1935), 347–402.

[1] LIAPOUNOFF, A. Sur une proposition de la théorie des probabilités. *Bull. Acad. Sci. St-Pétersbourg,* (5), 13 (1900), 359–386.

[2] LIAPOUNOFF, A. Nouvelle forme du théorème sur la limite de probabilité. *Mém. Acad. Sci. St-Pétersbourg,* (8), 12 (1901), No. 5.

[1] LINDEBERG, J. W. Eine neue Herleitung des Exponentialgesetzes in der Wahrscheinlichkeitsrechnung. *Math. Zeitschrift,* 15 (1922), 211–225.

[1] LUNDBERG, F. Über die Theorie der Rückversicherung. *Verhandl.* 6. *intern. Kongr. Vers.-Wiss., Wien,* 1909, 1, 877–955.

[2] LUNDBERG, F. *Försäkringsteknisk riskutjämning,* 1–2. Stockholm, 1926–28.

[1] V. MISES, R. Fundamentalsätze der Wahrscheinlichkeitsrechnung. *Math. Zeitschrift,* 4 (1919), 1–97.

[2] V. MISES, R. Grundlagen der Wahrscheinlichkeitsrechnung. *Math. Zeitschrift,* 5 (1919), 52–99.

[3] V. MISES, R. *Wahrscheinlichkeitsrechnung.* Leipzig-Wien, 1931.

[1] PEARSON, K. Historical note on the origin of the normal curve of errors. *Biometrika,* 16 (1924), 402–404.

[1] PÓLYA, G. Herleitung des Gauss'schen Gesetzes aus einer Funktionalgleichung. *Math. Zeitschrift,* 18 (1923), 96–108.

[1] RADON, J. Theorie und Anwendung der absolut additiven Mengenfunktionen. *Sitzungsber. Akad. Wien,* 122 (1913), 1295–1438.

[1] RIDER, P. R. A survey of the theory of small samples. *Annals of Math.* (2), 31 (1930), 577–628.

[1] ROMANOVSKY, V. Sur un théorème limite du calcul des probabilités. *Rec. Soc. Math. Moscau,* 36 (1929), 36–64.

[1] SLUTSKY, E. Über stochastische Asymptoten und Grenzwerte. *Metron,* 5 (1925), 1–90.

[1] "STUDENT". The probable error of a mean. *Biometrika,* 6 (1908–9), 1–25.

[1] THIELE, T. N. *Theory of observations.* London, 1903.

[1] TITCHMARSH, E. C. *The theory of functions.* Oxford, 1932.

[1] TODHUNTER, I. *A history of the mathematical theory of probability.* Cambridge-London, 1865.

[1] TORNIER, E. *Wahrscheinlichkeitsrechnung.* Leipzig-Berlin, 1936.

[1] DE LA VALLÉE POUSSIN, C. *Intégrales de Lebesgue, fonctions d'ensembles, classes de Baire.* Second ed. Paris, 1934.

[1] WINTNER, A. On the addition of independent distributions. *Amer. Journ. Math.* 56 (1934), 8–16.

SOME RECENT WORKS ON MATHEMATICAL PROBABILITY

BERRY, A. C. The Accuracy of the Gaussian Approximation to the Sum of Independent Variables. *Trans. Amer. Math. Soc.* **49**, 1941, 122–136.

CHUNG, K. L. *A Course in Probability Theory.* New York, 1968.

DOOB, J. L. *Stochastic Processes.* New York, 1953.

ESSEEN, C. G. Fourier Analysis of Distribution Functions. *Acta Mathematica* **77**, 1945, 1–125.

FELLER, W. *An Introduction to Probability Theory and its Applications.* Vol. I, 3rd ed. New York, 1968; Vol. II, New York, 1966.

GNEDENKO, B. V. and KOLMOGOROFF, A. N. *Limit Distributions for Sums of Independent Random Variables.* (Translated from the Russian by K. L. Chung.) Cambridge, Mass., 1954.

LOÈVE, M. *Probability Theory.* 3rd ed. Princeton, 1963.

LUKACS, E. *Characteristic Functions.* London, 1960.

ZOLOTAREV, V. M. A Sharpening of the Inequality of Berry–Esseen. *Zeitschr. Wahrscheinl.-theorie u. Verw. Geb.* **8**, 1967, 332–342.